The Limits of Reason

George Lowell Tollefson

PALO FLECHADO PRESS

The Limits of Reason
© 2020 George Lowell Tollefson

All Rights reserved.

ISBN-13: 978-1-952026-02-7

Library of Congress Control Number: 2020917149

Palo Flechado Press, Santa Fe, NM

**OTHER PHILOSOPHICAL WORKS BY
GEORGE LOWELL TOLLEFSON**

The Immaterial Structure of Human Experience
A Healer of Nations
Unbridled Democracy

Extracts from *Unbridled Democracy*

Spirit as Universal Consciousness
The Thinking Arts
Ethical Considerations
Moral Democracy

CONTENTS

Preface i
Ideal Concepts 1
The Right Angle 21
Logical Implication 41
Squaring the Circle 46
Euclid's Triangle 56
The 5th Postulate 58
Parallel Lines 66
The Angle's Origin 69
The Number System 72
Numbers 83
Unity 87
The Number Line 88
Proportion 93
Prime Numbers 95
Composite Numbers 99
The Infinitude of Primes 104
Mental Focus 108
The Binary Mind 115
Probability 118
Commensurable Relations 123
"Infinite" Sets 153
Science and Philosophy 158

Energy and Change 161
Space, Time, and Motion 165
Entropy 171
Order 174
Microphysical and Macrophysical 179
The Conceptual Reduction 183
Gravitational Force 188
Zeno's Paradox 191
Momentum and Location 197
Universal Complementarity 200
The Hidden Dynamic 203
Instrumental Science 208
Final Cause 210
The Discrete Mind 213
Index of Names 217
Bibliography 219

Preface

This book follows one titled *The Immaterial Structure of Human Experience*. Though an extension of that work, it makes little direct reference to its point of view. Rather, it is written almost entirely in terms of a representationalist examination of the mind's operations. It is limited to this viewpoint to avoid reference to the specialized terminology of the previous book. But it does not disagree with its immaterialist perspective.

What is of interest here is not the overall structure of human experience, its character, and its origin, but rather the limits of reason. This is forthrightly demonstrated by an insistence upon the fact that the foundation of mathematical science is empirical. However, this empirical influence is only initial. For mathematics is subsequently drawn into abstraction, where any further buildup of relations is imaginative and logical. But it cannot be overemphasized that the science begins with physical experience.

In keeping with its principle topic, the present work is replete with mathematical references. But its arguments and descriptions are not mathematical. There is no attempt to reason by proofs. And the mathematical concepts employed are elementary in character. What is presented is a general discussion aimed at demonstrating the imaginative origin of these ideas.

The examples which are given need not represent the path actually taken to invent mathematical concepts. They need only show that human imagination, working with the materials of ordinary experience, is all that was needed to erect the imposing edifice of this science. In sum, there need not have been, and indeed are not, innate ideas present in the human mind which transcend material experience.

It is more readily to be observed that physical science is a human creation and not something which might be thought to transcend physical experience in its fundamental principles. Yet, while it adheres to an observational organization of that experience, there is much that is not physical which is built upon the original findings. In addition, it should be noted that there are other aspects of human experience which do not fall under the purview of the scientific method of investigation.

For the above reasons, the present work develops a line of thought which does not accept any form of innate ideas. Nor can it tolerate an approach to the mathematical or physical sciences which suggests an idolatrous attitude toward them. These disciplines are in their entirety human inventions.

They are brilliant inventions. And their practical efficacy is acknowledged with due reverence. For they are viewed instrumentally as powerful tools. However, it is also understood that they are lacking in an access to final truths. These must be approached, if perhaps never fully realized, by other means.

The Limits of Reason

George Lowell Tollefson

Ideal Concepts

Mathematical reasoning provides a good example of the strengths and limits of the human mind. Take geometry, which underlies much of the thinking in physical science. Prior to the nineteenth century, this geometry was essentially Greek in origin. So it might be asked: if geometry underlies a science which attempts to describe the physical world, how accurately do its geometrical figures represent physical reality? And, by extension, how closely does mathematics in general reflect physical reality?

Geometrical figures are concepts in mathematical science. As such, they are formed in the mind. So what is the relationship of these concepts to the data of what are generally thought of as the senses? Are such concepts concerning the world exact? Or are they inevitably approximations?

To understand this, a closer look should be taken at the relationship between the human mind's powers of conceptualization and the things this mind is attempting to understand with such concepts. For the concepts, several examples from Euclid's *Elements* will do: the circle, the arc, the square, the lines which compose them all, and the angles which make up the square.

Pi is integral to an understanding of the ideal circle. Yet it is an irrational number. It is irrational because it expresses a relationship between two incompatible idealizations which lie within the definition of a circle. These are a fixed radius and the resulting circumference. By an idealization is meant a concept which is created in the mind, rather than from an image which is formed by physical experience. The former can be distinguished

The Limits of Reason

from the latter by the fact that its existence as a concept is unique to thought and not found in nature.

The circumference, or uniformly continuous arc, of a circle is such a concept. It is defined by Euclid as equidistantly surrounding a center point. This is determined by a fixed radius.[1] If it should be postulated that this might occur in nature, it cannot be demonstrated that it does. Another similar ideal concept is that of a straight line of fixed length, as in the radius of that circle. Are there any perfectly straight lines in physical experience? And are there multiples lines of exactly the same length? And, again referring to the circumference, how many such radii would be needed to guarantee that it is a uniformly continuous arc?

These concepts stand in contrast to one which is formed directly from the experience of mental impressions like hard, cold, round, and white. These latter impressions are associated together in an image, a snowball, which is considered concrete in character, since the type of the impressions and the manner of their association are directly encountered in physical experience.

When left unaltered, it is this image which supports a concept that is true to that experience. Note that a physical concept such as this tends to be simpler than the ideal concepts just mentioned. That is to say, it is immediately familiar. But this simplicity does not mean that it cannot be a compound of multiple impressions, just as the simplest of the above ideal concepts is.

---

[1] *Euclid's Elements*, Book I, Definition 15.

For a deceptively simple example of idealization, the definition of a line is given as a "breadthless length."[2] Numerous impressions make up any image of breadth or length. And the idea of something which is breadthless involves a negation as well: a conceiving of breadth accompanied by a denial of its application to this case.

A straight line is another example which is even more ideal in character. For it is defined as "a line which lies evenly with the points on itself."[3] In other words, if the points of this line are marked upon a plane, and the line is removed then returned to match up with at least two of those points, it will match up with any of the rest of those points which lie along its length. This differs from a curved line, which, under the same circumstances, can only make contact with all those of its points in the plane which are covered by its length if it remains in the initial position in which these points were originally laid down.

But in the case of a concrete image matters are different. Since it represents perceptual experience, it comes from nature. For example, the mental image of a snowball arises from a compound of associated impressions which are physical in character. It is a physical object which is hard, cold, round, and white. This is what the image formed in the mind conveys.

The concept of a snowball derived from that image is not markedly different from the image. This is the case so long as such characteristics as the roundness of the snowball remain understood as having been derived from physical experience and not as having been idealized by further embellishments of the imagination. In other words, the initial image will retain its

---

[2] *Euclid's Elements*, Book I, Definition 2.
[3] *Euclid's Elements*, Book I, Definition 4.

The Limits of Reason

concrete character, until someone deliberately begins to compare it to images drawn from other similar objects and thereby forms a more generalized concept of all of them.

From that point he is likely to proceed to incorporate the associated properties which characterize the object, as well as those which do not, into a more refined definition which no longer accords directly with physical experience. This type of abstraction is well illustrated by the concepts of a perfect circle and a breadthless line. For the perfect circle is derived from multiple instances of less than perfect circles in experience. And the breadthless line is idealized from a line with breadth, which breadth is necessary for it to be physically experienced.

So the simple definition of a snowball references physical experience. And it is understood to be a product of certain conditions such as winter, cold temperatures, and human bodily agency in forming its texture and shape, all of which are elements of physical experience. It is these, along with the light reflective properties of the ice forming the snowball, which make it hard, cold, round, and white.

Now, once again, there are two ideal geometrical concepts which are integral to the definition of a Euclidean circle. They are a continuous, unvarying arc comprising the circumference of that circle and an unvarying straight line of specific length comprising the radius of that circle. These two concepts are not only ideal. They are mutually exclusive of one another.

This is because they are independent creations of the mind. They are not either of them found in physical experience. Such experience might be supposed to supply a connection between them. But it does not. This is because any physically encountered arcs, lines, and the circles they compose are not Euclidean.

They lack the precision and uniformity—one might say, the artificiality—of his definitions.

So the sole connection between these two initial Euclidean concepts—the unvarying arc and the straight line of a specific length, which are the circumference and the radius of a perfect circle—lies in the fact that they are both ideal concepts. Ideality is the only thing they have in common.

It is true that ideal concepts like the uniform arc and the straight line of a specific length, or other concepts derived from them like the circle, may be logically connected to each other or to other such concepts in a train of thought. They may even have been built up into an entire system of logically related mathematical propositions, as in Euclid's *Elements*. But they are otherwise mutually exclusive. There is no connection between them, other than that of the careful relating of propositional terms which constitutes logic.

Note, for instance, that the perfect circle is offered as a definition and not as a theorem in the *Elements*. It does not need demonstration, or proof, and is thus given an independent status, like that of an axiom. But, in spite of the fact that it appears to be a self-evident concept, it is in truth a composite product of imagination.

After Archimedes' work on getting a reasonably accurate, but inexact, number for pi,[4] it has been made ever more clear that the connection between the circumference and the radius (or the diameter) is not strictly quantitative. Otherwise, today the formula for the Euclidean circle would not be $C = \pi d$ (or $C = 2\pi r$), in which π is an irrational number. So, looking back to Euclid, it can be seen that the concept of a circle is clearly an

---

[4] Archimedes, "Measurement of a Circle," Proposition 3.

The Limits of Reason

imaginative invention which is more complex than that employed to define a straight line.

Since it can be asserted that the circumference and radius of a perfect circle are imaginative idealizations with no connection between them but one of imaginative invention, it is clear that they cannot have a connection between them in physical experience. In fact, their ideal character implies that they are themselves more generalized, and thus more highly abstracted, than any concept based directly on physical experience.[5]

This creative contribution of the mind is what it means to develop an idealization, let alone further developing a more complex idealization from the initial idealizations. Consequently, the perfect circle, which is derived from the concepts of the unvarying arc and multiple straight lines of a specific length, can only be an approximation of what is found in nature.

No one has ever seen a circle with a perfectly regular circumference. Nor has anyone seen a perfectly straight line, nor a multiple of lines of the exact same length. Nor, following Euclid's definition, has anyone seen a single-dimensional line, for that matter. If it is the case that a perfect circle, or a perfectly straight line, or a multiple of lines of the same length, have ever been encountered in experience, it was not known at the time. Or, at best, one could not be certain of the fact. And certainly a breadthless line can neither be encountered nor imagined.

It is this idealization and physical unreality which renders any two geometrical concepts incompatible in their relationship

---

[5] As to a concept based directly on physical experience, it is also abstract in the sense that it is represented by an image in the mind which does not incorporate all the details from the actual experience. But there is no imaginative invention of those details. There is only a selection of them.

to one another. Since they have no connection in nature, they are utterly distinct and separate idealizations. Hence the existence of pi, the irrational constant which imaginatively connects an unvarying arc and a straight line in an ideal circle. The indeterminate number represented by pi reflects the uncertainty of the relationship in spite of its imaginative invention. It is an unquantifiable relationship.

The justification for saying the relationship is unquantifiable is to be found in Euclid's definition for an ideal circle, which is stated as:

> A circle is a plane figure contained by one line such that *all the straight lines falling upon it* from one point among those lying within the figure are *equal to one another*.[6] [The italics are added.]

So, to simplify this discussion, let it be said that the straight line—the radius of the circle—is determined at a measure of 1/2. That would make the diameter 1, and thus, using the formula πd, a measure of the circumference which is π is arrived at. Pi is an irrational number, which is indeterminate. Thus the circle's circumference is indeterminate in this case.

How is this so? It is so because the relationship between the circumference and the radius is by definition always indeterminate. This is because the ideal perfect circle is defined in terms of *all* the radii falling upon its circumference. The "all" is an indeterminate number of radii, which in turn implies an indeterminate number of points of contact of those radii with the

---

[6] *Euclid's Elements*, Book I, Definition 15.

The Limits of Reason

circumference. Thus the circle's circumference cannot be computed in terms of a rational number.

Why is this? Imagine a kink in this circumference, a sudden angular change of direction in the arc. Such a bend might be microscopically small. It could be indeterminately small. Yet the bend would contradict Euclid's original definition of a circle. Thus an indeterminate number of radii, and a correspondingly indeterminate number of points in which the radii make contact with the circumference, must be postulated for the definition to hold. This is precisely what the indeterminate, or irrational, number π represents, insofar as it is an irrational number.

But again why, it might be asked, is it possible to have a determinate circumference, which would then require an indeterminate diameter. For instance, this would be the case if the diameter were $1/\pi$. That would make the circumference 1. This occurs because incompatible concepts are being related, one determinate, the other indeterminate.

So, if one of them is assumed to be definitively rational—i.e., determinate—this renders the other indeterminate, or irrational, because there is something indeterminate in the relationship between them. If the diameter is determinate, the circumference is not. If the circumference is determinate, the diameter is not. Hence pi, the constant of indeterminacy which stands as the relationship between them.

It follows from this logical but quantitatively inexact relation that the Euclidean circle is an idealization brought about by the union of two mutually exclusive ideals. In Euclid's *Elements* this is accomplished by fiat. It is set down as an initial definition,[7] thus avoiding the inconvenience of a logical proof. Why the avoidance is not clear. Perhaps the circle seemed too obvious to be a concept which required demonstration. At any rate, its status as a definition obscures its distance from nature.

But let this discussion not stop with the circle. To get a sense of how pervasive this problem is, let a near relation to it be examined: the ideal square. Here too an irrational number will be discovered. It is the square root of two. If a square is constructed with sides of one unit length, by means of the Pythagorean theorem a diagonal which is the square root of two will be obtained. This is very much like pi, inasmuch as it is an irrational number.

The square root of two is always involved in the relationship between the sides and diagonal of a square: $d^2 = 2a^2$, "d" being the diagonal and "a" each of two adjacent sides of the square. Therefore $d = a\sqrt{2}$ or $a = d/\sqrt{2}$. Thus, if a, or each of the sides, is $\sqrt{2}$, the diagonal can be rational. It would be 2. But if the sides of the square are a rational number, the diagonal is irrational.

Now it is clear in some cases that, if the sides of the square are an irrational number other than $\sqrt{2}$—say they are $\sqrt{3}$—then both the sides and the diagonal can be be irrational at the same time. However, it will not be the case that the sides of the square are rational and the diagonal is also rational.[8]

So there is an irrational factor, namely $\sqrt{2}$, always at play in the relationship between the diagonal and the sides of the square. This presence of an irrational factor in the relationship

---

[7] Ibid.
[8] It is also true that both the diameter and the circumference of a circle can be irrational together, but not both rational together.

The Limits of Reason

between the diagonal and sides of a square is similar to that which is found between the diameter and circumference of a circle.

So it should not be altogether surprising that a square can be inscribed within a circle. And, when it is, its diagonal will become the diameter of the circle.

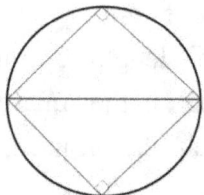

As a diameter, this line will always be irrational in its relationship with the circumference of the circumscribing circle, just as the diagonal is irrational in relation to the sides of the square. For, in the case of the circle, the relationship will be determined by the constant pi, just as, in the case of the square, it is determined by the $\sqrt{2}$.

Furthermore, the four corners of a square are right angles. Thus the perimeter of a square encloses an area of 360°, just as does the circumference of the circle that circumscribes it. Though it is true that their areas are different, their bisecting lines are the same. The diagonal and diameter are equivalent. Is it any wonder then that such a diagonal might be represented by an irrational number, either in itself or in its relation to the sides of its square? It certainly has this relationship to the circumference of the circle which circumscribes it.

If the diagonal were a rational number, the sides of the square would not be so. This is simply a result of a transference in the relationship. The square root of two continues to appear either in the sides of the square or in its diagonal. There is always an irrationality in the relationship. This is similar to the relationship between the circumference and diameter of a circle.

It can easily be seen that it is the relationship of the diagonal of the square to the circumference of the circle that is of concern. In other words, what matters is the fact that a square can be inscribed within a circle, thus making its diagonal the diameter of the circumscribing circle. As a result, when relating the sides of the square to its diagonal or the circumference of the circle to its diameter, there is an irrationality in the relationship.

But how is this irrationality caused? It is caused by an indeterminacy which is to be found both in the ideal square and in the perfect circle. This problem of indeterminacy occurs in the perfect circle because what is being attempted is to determine the uniform curvature of an arc by means of an indeterminate plenitude of equivalent straight lines.

The circumference is indirectly defined by Euclid as a perfectly unvarying curve. For, if "all the straight lines falling upon [the circumference] from one point among those lying within the figure are equal to one another,"[9] then the circumference is a perfectly unvarying curve. The words "perfectly unvarying" may not be his. But they are his meaning.

Can a perfectly unvarying curve be physically demonstrated? It cannot because it is an ideal. It is an ideal abstraction from physical reality. The ideal abstraction has conceptual validity as a thought. And the varying curve of a natural circle (as opposed to "perfectly unvarying") has physical validity as something in nature. But an ideal thought is not physical. It is a free product of imagination, which rearranges the perceptual content of

---

[9] *Euclid's Elements*, Book I, Definition 15.

The Limits of Reason

physical experience. So the leap between physical experience and the ideal cannot form a determined connection.

The same thing occurs with a square because the concept of a square is ideal. It is not found in nature. What specifically is not found is a perfect right angle. Nor can four absolutely equivalent right angles covering the full revolution of changes in direction of the circle's circumference be found.

Nor can four absolutely equivalent straight lines be found. If any of these things were found, they could not be identified. So the square, composed of four right angles and four absolutely equivalent lines, is one idealization set among the many natural quadrilaterals to be encountered in the world, none of them precisely measurable.

It is in this way that the irrational character of both the numbers under present consideration (pi and the square root of two) reflects the ideality of the perfect circle and the square. An irrational number is specifically an indeterminate number. And there is something indeterminate in the ideal composition of both the circle and the square. But the question remains: why is the relationship between the diagonal and sides of a square indeterminate?

Of course, it is a relationship between two ideal concepts: that of a perfectly straight diagonal line and that of the square's perimeter with its precisely indicated angles and straight lines of equal length. But this sort of thing is true of all representations of geometrical figures. Any ideal geometrical representation is potentially incompatible in the relation of its parts. For the parts are individually ideal. Thus their quantitative measures and the relationships between them are assumed ideally.

But most of these figures do not exhibit relations which are assumed to be incommensurate by definition. In other words, an incommensurate relationship between the parts of a geometrical figure is generally not explicitly stated in the manner in which it is in Euclid's definition of a circle. A perfect circle clearly exhibits an incommensurate relationship, when its definition specifies that the number of radii needed to secure an unvaryingly uniform circumference is indeterminate.

So there must be something to further explain the similar relationship between the diagonal and sides of a square. And that something is the fact that the relationship between the diagonal and sides of a square yields an irrational number. It does so because the character of the square results from the nature of its enclosure in the circle which circumscribes it. It results in the square root of two being directly linked to pi.

This link is initially foreshadowed by the equivalence of the diagonal of the square and the diameter of its circumscribing circle. But, in addition to this, the circumference of the circle is proportionately linked to the four equal sides of the circumscribed square. For the sides act as chords apportioning the circumference into four equal arcs. In turn, the four equal arcs together comprise the full circumference of the circle. Thus, as the relationship between the diameter and circumference of the circle is incommensurate, so the relationship between the diagonal and perimeter of the square is also incommensurate.[10]

---

[10] A further observation is that the algebraic root, the square root of two, is proportionately related to the transcendental number, pi. For a successive bisection of the sides of the square to a circumscribed eight, sixteen, thirty-two, etc. would bring the square's perimeter progressively closer to the circle's circumference. Thus the relations of these to their respective diagonal or diameter would become more alike. Even so, an exact numerical equivalency cannot be obtained.

The Limits of Reason

As already indicated, all geometrical figures are ideal, though they may or may not involve indeterminate relations which are definitionally specified. Nonetheless, the ideal character of their internal relations does imply an uncertainty in the relations of their component elements. For, as previously stated, regardless of the geometrical figure it belongs to, the exact measure of a line is unknown in physical experience. Nor can there be an exact sense of the measure of any angle in physical experience.

The precise character of an angle cannot be physically known because that character is ideal. It is derived by logical deduction from the circumference of a circle, or by a revolution through a Cartesian plane. One complete revolution is 360°. Thus a 90° angle is one fourth of 360°, etc.

Trigonometric functions might appear to contradict this claim about the vagueness of angles because they express angles in terms of a ratio of lines—i.e., the sides of a right triangle. But that merely transfers the problem to the indeterminacy of lines. Moreover, most of these trigonometric values are found to be irrational (or indeterminate) due to the commonly incommensurate character of the sides of a triangle, which only further supports the thesis of ideality.

So, if angles are to be referenced to a circumference in some manner other than an arbitrary division into degrees, the proportions of that circumference can only be determined by the circumference's relationship to the circle's radius. Hence the use of radian measure. However, this measure is indeterminate, as it involves pi. So there is a numerical impasse in working out such geometrical concepts, should they be supposed to reflect exact quantitative relations.

In general, it would appear that geometrical concepts belong to their own logical and conceptual reality. Thus they do not arise directly from the physical world. But they are certainly derived from it indirectly. They are idealizations and regularizations of what can be found in irregular plenitude in the physical realm. If mathematical concepts were not thus operatively connected with the physical world, they could not reflect definitive physical results.

Nevertheless, the fact remains that the concepts are ideal. For conceptual thought is made independent of any mental representation of perception through a process of abstraction from nature. And the most highly abstracted type of concept is ideal. It is therefore not directly, but only indirectly, representative of nature.

So, following the implications of this analysis, similar arguments may extend beyond the circle and the square to other geometrical figures, insofar as the ideal nature of such figures is concerned. However, what has been unique to the square and the circle is the definitionally irrational character of the relationship between certain of their constituent parts. This is not the case in a figure like a triangle. For an indeterminate relationship between its components is not definitionally alluded to.

Now let the square's symmetrical relationship to its circumscribing circle be looked at again. A square is proportionately linked to this circle in both its angles and its sides, not to mention the identity between its diagonal and the circle's diameter. It is this close multiple relationship which makes a square and a circle so similar in terms of the irrational relationship between their bisecting lines and their perimeters.

Any polygonal figure described by straight lines must be defined by angles. Both the angles and the straight lines would be

The Limits of Reason

physically imprecise in character if sought out in experience. It is for this reason that they are considered to be ideal. An attempt to relate angles to lines with quantitative precision is a purely intellectual activity. All of Euclid's polygons are therefore ideal, though this may not be as apparent as it is in the case of the circle and the square, where one is confronted with a definitionally specified quantitative dissociation of component concepts.

The point in presenting this discussion of arcs, circles, lines, angles, squares, and polygons in general has been to emphasize that they are all idealizations. This can be demonstrated for all of Euclid's geometrical figures, even if their ideal character is more subtle and must be analyzed in ways other than that of the circle and the square. For example, as can be seen, a rectangle is not readily susceptible to an analysis resembling that of the circle and square. But it does have ideal components, such as its four right angles and its straight lines of specific lengths.

The concept of angles is derived from the circumference of the circle, which is ideal in definition. The lines which constitute the sides of a rectangle are also ideal in definition. They are thus conceptually separate from angles, insofar as their abstraction from the physical world is concerned. They are concepts formed independently from, and without immediate regard to, the concepts of angles. Though, of course, it would not be feasible to execute a graphic representation of an angle without lines.

So angles and lines are not only independent of one another in their initial development as concepts. They are incompatible in a certain sense. For, as ideal abstractions independent of one another, they have no connection to one another in nature. To form a rectangle, they must be brought into union by an increased process of idealization as a result of an imaginative connection being made between them.

For example, it is an intellectual process which informs a person she has encountered an angle in a cliff or shoreline. The concept "angle" does not correspond directly to natural phenomena. Such a concept entails a precision regarding one angle in relation to other angles, which is a product of thought, however primitive in development that thought may be. Without some sort of mental processing, there would simply be a sense of proximity and recession in physical experience. It is the intellectual faculty of the mind which refines primitive experience by means of the development of an overlay of such precision.

So the kinds of geometrical angles and lines seen in Euclid are put together by imagination. In other words, a new geometrical concept, such as a Euclidean triangle or rectangle, which is hewn from the definitions of angles and lines, must be built in the mind without further reference to nature.

That new concept, say the rectangle, is intellectually, not experientially, derived from its constituent ideal parts, the line and the angle. Of course, imperfect trilaterals and quadrilaterals, which resemble triangles and rectangles, are encountered in the physical realm. But they lack the ideal relations of these figures.

All geometrical figures were intended by Euclid and his predecessors to represent the physical world. And they were thought to do so until fairly recent times. Yet these geometrical figures of the mind do not now, nor ever did, exist definitively within the physical world. Rather, it must be said that they have been employed to map it.

But no matter how close to physical experience their representations may be thought to be, however refined they are, they are not exact equivalents of their physical counterparts. They

The Limits of Reason

are conceptual approximations of them. Thus a good terrain map will carry a military patrol through unfamiliar country. But the map could not be imagined to stand in for all the features, visible and hidden, of that country. Nor are the map's terrain features exact equivalents of the few selected physical forms they are intended to represent.

Overall, what this discontinuity between abstract concepts points to is the simple fact that human reason—perhaps especially in its most impressive idealizations, such as the square and the circle—is a fallible instrument. It is makeshift. Needless to say, it works very well in the investigation and control of specific aspects of nature. But what works is not necessarily what is true in an absolute sense.

Mathematical theorems and scientific theories are conceptual and logical in character, since they inevitably involve idealizations and associations between those idealizations. The initial images are representative of objects in the physical world which are refined by imagination (i.e., made simpler and more regular). In this way, each of them is converted into a concept by the precise application of a definition peculiar to its properties.

In the case of mathematics, this is why the elements which contribute to a geometrical theorem must be conceptualized as definitions, and the relations between them as axioms. Subsequent to this initial conceptualization, geometrical figures can be logically (which is to say, associatively) derived from such definitions and axioms, and from other figures already so derived. For the whole of a theorem is abstractly conceptual, both the derived proposition and its constituent terms.

Euclid's definitions and axioms provide a foundation for his propositions in just this way. The theorems are determined by the logical structure of mathematical science. The logical structure is the set of rules by means of which prior idealizations are put together to create new ones. The rules, which are conceived to be consistent among themselves, determine specified relations of association.

In other words, specific rules apply according to regularized methods of association, as opposed to other rules which might employ other possible methods of association. It is in this way that in arithmetic the assumed equivalence of arithmetical units and the character of the number line are used to determine the numbers 12, 7, and 5, and then, by extended use of these same preliminary assumptions, to further determine that 12 minus 5 should equal 7.

Likewise, a perfect circle must be composed of multiple equal straight lines of a specific length and of a circumference of unvarying curvature because this logical compounding of these prior abstract concepts is how it is defined. The equal straight lines and the unvarying curve are brought into an association with one another by the definition. But none of this gives any of these ideal concepts a special existence either in nature or in some transcendent spiritual realm. They are imaginative idealizations originally derived from abstractions taken from nature.

It is the same with physical science. Experimental verifications of observed phenomena are dependent upon the structure of a theory and the perspective it supplies. Any such theory is logical in character and made up of ideal concepts. It is therefore abstract. A theory has initial components, which are its definitions and axiomatic assumptions. These components and their logical derivatives generate a hypothesis for a further development of the theory. It is this hypothesis which must be tested to get affirmative or negative results.

The Limits of Reason

In fact, the preliminary situation must begin with concepts of the natural components brought under observation. Then the results of experiment must be conceptualized as relations of the observed components. All the concepts and relations will be in some degree ideal, even the most experiential of them. For they are mental constructs conditioned by a theoretical perspective. To the extent that these things are done, a rational structure is established which is parallel, but not equivalent, to nature. This is the experimentally confirmed theory.

For example, if the charge and mass of an electron are measured, as Sir J. J. Thomson did, is it truly known what has been measured? A person does have a conceptual grasp of what has been done. But the physical ground of that research is less certain. For, as regards the relationship between the conceptual and the physical, there must always remain some doubt.

Insofar as any of these concepts, relations, and theories involve idealizations—that is, inasmuch as they involve thoughts that are refined abstractions from physical experience—they are only productive of approximations of what is experienced in the physical world. They are like terrain maps, very sophisticated templates placed over the inscrutable whole of physical reality. Science can get where it is going only because these maps are good enough to get it there, not because science has the fullness of reality in its grasp.

George Lowell Tollefson

The Right Angle

To begin, let a quadrilateral be imagined. It is desired that this particular quadrilateral should have equal sides. So this will be assumed. And the quadrilateral will be called a rhombus. In addition, the sides of this rhombus will be set perpendicular to one another. This is to say that each side will extend from another in such a manner that, if the base line were extended beyond both sides of its adjoining line, the adjoining line would be no nearer to the base line on one side than on the other.

It can now be noted that, when straight lines are drawn between vertices which are not constructed from the same sides of this quadrilateral, the lines will intersect. And the point of intersection will appear to be equidistant from the four sides. The resulting figure will be called a square. And the intersecting lines will be called diagonals. They are being given these names because they have just been discovered in the manner described.

Furthermore, it can be seen that there are four angles between the diagonal lines at the point of their intersection. Accordingly, there is one of these angles in each of four three-sided parts which together make up the quadrilateral. Let the angles be called central angles. And let the parts be called triangles.

Having done these things, the mind has imaginatively created a square. But it does not yet know conceptually what constitutes a square, other than that the figure has more or less equal sides which are more or less perpendicular to one another. For it is hard to get things exact in a mental picture or a drawing. So let such a figure continue to be imagined and see what is found.

The Limits of Reason

Let an attempt be made to inscribe that square within a circle:

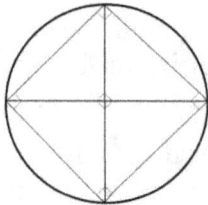

It seems to work in a pictorial sort of way. Remember, the intellect does not yet conceive that the square has right angles. For a right angle can only be a concept: a concept being a set of images representing properties and limited by a definition. Nor does the intellect understand that the drawn circle circumscribing the square has equal radii. For the square is an imaginative figure within another imaginative figure, the circle. Neither is conceptually derived at this point.

Again, it should be noted that the term "perpendicular" indicates a straight line extended away from another straight line, keeping itself as equidistant from that line on either of its sides as it can. So the imagination has not yet defined a right angle. For a right angle is a conceptual abstraction, the correlative of which is not found (or at least not confirmed) in experience. It is an idealization.

This is to say that, when Euclid offers such a definition, he states that the "right angles" on either side of the perpendicular line are equal.[11] But, once defined in such a manner, the term "right angle" becomes a concept, particularly when it is univer-

---

[11] *Euclid's Elements,* Book I, Definition 10.

salized by the fourth postulate, which specifies that "all right angles are equal to one another."[12] What was previously only pictorially supposed is made certain by definition. However, the present discourse has not reached the conceptual stage.

Of course, straight lines that are truly straight, perfect circles, and equal entities are all conceptual idealizations, insofar as they are so defined. But they can also be imagined without taking the extra step of conceptual definition. However, this is at some cost of imprecision. For, as the imagination shifts in its representation of something, the undefined image in its grasp is subject to subtle or greater changes. Thus it is flexible and imprecise. So a definition is required to set the image into conceptual exactness.

It is in exception to this limiting circumstance that the imagined figure above is being called a square. It is not a defined concept at this point. Rather, it is proposed to be a square simply because it is four-sided and the four sides appear to be equal and perpendicular to one another. The exactness of these features has not been determined.

But having drawn lines as straight as possible from corner to corner of this square, what begins to emerge (as can be seen in the figure above) is a recognition that, where the diagonals intersect, the vertices of the central angles formed by this intersection appear to meet at or near the center of the circumscribing circle. In other words, the two diagonals seem to bisect one another.

And, if the diagonals are the diameters of an imagined circle, such as the one circumscribing this square, they are also roughly equal to one another. For there should only be one common

---

[12] *Euclid's Elements,* Book I.

The Limits of Reason

measure for the diameter. Otherwise, the circle would be an ellipse or some visibly lopsided figure. This can be demonstrated by graphic experiment.

Draw any circle. It will be seen that the more irregular the lengths of the diameters are in appearance, the more irregular in curvature the circumference is. So, if these observations are accepted hypothetically to be associated relations of likeness (what would by definition be called equality), then, on that basis, it can be further asserted that each of the four interior triangles of the square have roughly two equal sides which enclose each central angle at the intersection of the diagonals.

This assertion would no doubt appear to be logically inferred. But, in fact, it is simply the mind making a more complex association between two elemental associations. This would correspond roughly to Euclid's first common notion:

> Things which are equal to the same thing are also equal to one another."[13]

Euclid's statement is, of course, logically rather than loosely associative. Though its logic is not demonstrated. It is assumed axiomatically.

So, remaining short of a logical inference, the two diagonals of the imaginative image can only be said to be apparently equal. Likewise, an apparently bisected diagonal would be divided into roughly equal parts. And those roughly equal parts would seem to be roughly equal to the parts of the other diagonal.

---

[13] *Euclid's Elements,* Book I, Common Notion 1.

An association such as this is, of course, what a logical inference is. But an inference can only be derived from reliable (i.e., stabilized) concepts. With imaginative images, it remains a loose association and no more. So, for purposes of the present discussion, it is convenient to remember that this association is being made at the imaginative, not the conceptual, level.

It can now be speculatively assumed that the equal sides of the four triangles really do proceed from the center of the circle. Again, this is the simplest of assumptions, which is a matter of imaginative association: an apparent equality of lines extending to the circumference of the circle. But it does not determine that the circumference itself is perfectly circular at all points. For, like the imagined square, the circumscribing circle is imagined, not ideal.

Nevertheless, it is noteworthy that all these circumstances can be visually recognized concerning the figure above. This can be done without resorting to any measurement, other than comparisons made with the eye. Nor are there any definitions which would require one to move between fixed concepts by means of logical inference.

So, although it is certainly true that the supposed equalities can neither be physically nor logically determined at this stage in the imaginative process, they certainly look the way the various assumptions have indicated. In fact, remembering that all four sides of the square appear to be equal, it is hard to imagine that the parts of the intersected diagonals would not also be equal. And that just by looking at the drawn figure.

This, however, is certainly not a proof. But there is a powerful incentive for accepting this pictorial thinking. Imaginatively, it looks that way. Of course, undemonstrated imagination is often proven wrong in mathematics. But proofs are not the object

The Limits of Reason

of this discussion at this point. Invention is the object. In other words, where did these geometrical concepts come from in the first place? How did the concepts originate well before it was understood how to link them logically?

Given the visual appearances, as exhibited in the figure above, it does not seem to be a great leap of imagination to assume without logical inference that the diagonals bisect each other, creating four equal lines radiating from the same point to the circumference of the circumscribing circle. So let this imaginative progress continue toward a point where the logical relations might have been established.

In accordance with Euclid's undemonstrated definition of an ideal circle, and in conjunction with his equally undemonstrated definition of the diameter of a circle,[14] equal straight lines proceeding from a central point within a circle to its circumference may be understood to be the circle's radii. Moreover, it should be recalled that the perfect circle has been discussed previously.

So its ideal character is now being accepted as it stands in Euclid's *Elements*. That is to say, it is accepted, but unproven. It is an ideal definition of a schema established by rule and compass. In other words, the drawing, which is an image, has been firmly set in character by a definition, rendering it a concept. However, nothing prior to its existence has demonstrated it as such.

It should be noted as well that each of the equal sides of the four circumscribed triangles is shared by two of the other triangles. Thus there are only four radii among the lines presently drawn within the circle. Because all these sides of the triangles

---

[14] *Euclid's Elements*, Book I, Definition 17.

are now accepted as being equal to one another, the possibility that the four adjacent triangles are congruent can be entertained.

Are not the four sides of the square presumed to be equal? These are the bases of the four triangles. That would make all three sides of any one of the triangles equal to the corresponding three sides of the other three, which, from a purely pictorial perspective, makes the triangles appear to be congruent. So the square must be made up of four congruent triangles, each of which has two equal sides. Let them be called isosceles triangles.

It is now fair to assume (not in strictly logical terms, of course) that the four central angles should be equal because the triangles are symmetrical to one another in appearance. If the triangles should be stacked on top of one another with the four central angles at the same end, which has already been suggested by their congruence, it seems likely that the central angles would be equal. For the base sides of the triangles would have to be made comparatively longer or shorter to render these apical angles unequal.

But, as has already been indicated in describing the square, the base sides of the four triangles are, in fact, equal. For they are the equal sides of the square. The same is true of all the base angles of the triangles. They are equal as well. Or at least it would seem that they should be, since the triangles appear to be congruent.

And, the two sides extending from the central angles being equal, the congruency would be the same no matter which face of the triangle is laid against another. So this is precisely what is desired of the square, that it is a quadrilateral composed of four equal parts—parts which are equal to one another in every way.

The Limits of Reason

The fact that the square's diagonals are equal can now be added with even greater assurance to its provisional character. For the diagonals are diameters of the circumscribing circle. And the diameters are assumed to be equal because each of them is composed of two equal radii proceeding from the center of the circle to its circumference.

That is Euclid's definition of the circle. But it cannot be a final definition. For a final definition would not only state, but suggest a means of demonstrating, the correctness of the properties involved. These properties have only been imagined and visually compared in this discussion. And they have been definitionally proposed without proof by Euclid. Neither an imaginative supposition nor an axiomatic approach will do for the purpose of accepting a final definition.

So, in addition to what had been previously stated, the square can now be divided into four equal parts, as has been indicated. Therefore, assuming the equality of the diagonals (which are diameters of the circle as well), and imagining as many diameters as may be desired, the regularity of the circle's circumference may be assumed.

Of course, this regularity is not known to be the case because imagination cannot supply a sufficient quantity of diameters to demonstrate that it is so. But, since all the diameters which can be imagined are properly circumscribed, the arc of the circumference making appropriate contact with the ends of each half diameter, or radius, it is at least pictorially probable that the circumference would form an unvarying arc.

Moreover, it can be seen that the equal bases of each triangle, which are equal sides of the square, are the sides of each triangle which subtend a central angle—i.e., an angle at the center of the circle and square. From this observation, and from the fact that these central angles are presumed to be equal, it can be assumed that the sides of the square demark equal portions of the circle's circumference.

So, since the lines which make up the sides of the square touch the circumference at both their limits, the sides of the square can be designated as chords. And the demarked portions of the circumference can be called arcs. There are four of these arcs, as there are four sides of the square. Let the areas between the arcs and their respective chords be called segments. They abut one another at the points where their chords touch the circumference. Thus the four arcs together express the full circumference of the circle.

Now, since the chords are equal and each of the arcs is presumed to be uniformly curved in the manner of the others, it is for this reason that the arcs would appear to be equal. But it must be granted that this assumption, as well as others which have been made, involves logical deduction: if such-and-such, then such-and-such.

Such thinking is inescapable. For it occurs in this case at a most fundamental level, which is that of Euclid's first and fourth common notions:

> Things which are equal to the same thing are also equal to one another.
> Things which coincide with one another are equal to one another.[15]

---

[15] *Euclid's Elements*, Book I, Common Notions 1 and 4 respectively.

The Limits of Reason

In fact, so fundamental are the common notions, even to imaginative thinking, the use of them appears frequently throughout this discussion.

This is because the common notions are direct associations. And direct associations are the work of imagination. They become principles of reason, or logic, only when defined as such. The simplest of mathematical operations are likewise directly associative in character when performed with small numbers, as can be seen in the next paragraph, where simple division is a matter of subtraction.

It is generally accepted that the circumference of any circle is arbitrarily designated to be 360°. The number 360 is even and therefore separable into equal parts. These equal parts are also even numbered. So they can be separated in the same manner again. Thus the entire operation follows the common notion:

> If equals be subtracted from equals, the remainders are equal."[16]

So, if the circumference is 360°, and the four lesser arcs are identical, then each of them must be a quarter of the whole, or 90°.

Since the central angles subtend equal bases of the triangles formed by them, and those bases are the chords subtending the arcs, the angles can be given the same measure as the arcs. So the angles subtending those arcs will be 90°. And a clockwise sweep of the four equal arcs of the circumference brings one back to a point of origin on that circumference.

---

[16] *Euclid's Elements*, Book I, Common Notion 3.

In this way, the central angles make up the whole of both the circle and the square, sectoring each of them in accordance with the four principal directions of a Euclidean plane: up, down, right, and left. So the angles will be called right angles. For a right angle represents half a rotation from one point on a straight line to another, thus rendering it perpendicular to the line. And, since in a two-dimensional plane there are two sides to a line, there are four half-rotations in one circuit about that line. Thus there are four right angles in the square. And there are four triangles, each with one right angle.

Now the concept whose origin is being sought is this right angle. And the observations which have been made suggest that it was imaginatively derived in a manner resembling the preceding discussion. But that is far too complex and delayed a process. The right angle was probably initially discovered as a quarter of a circular revolution of the human arms: beginning with one arm pointed upward, then pointing to the right side, then downward, then the other arm pointing to the left side, and upward again. Only subsequently was the angle recognized in terms of a quadrilateral like the square.

It is beginning to be noticeable that, though the area of the square is less than that of its circumscribing circle, as is clearly illustrated in the figure above, it exhibits properties similar to those of the circle. For example, the perimeter of the square is encompassed by four equal straight lines, each of which successively originates perpendicularly to the end point of a previous line and terminates perpendicularly to the origin of a subsequent line. This forms a single complete revolution which begins and ends at the same point. Such a revolution is also the case with the circumference of the circle, which curves uniformly from a point until it reaches that point again.

The Limits of Reason

But the perimeter is not a uniform curve like the circumference. The lines shift orientation by sudden turns, which form four angles. Otherwise the lines, being individually straight, would continue together on a straight course and would not encompass an area as the circle does. So, it may be asked, what is the measure of these angles which perform a function like the circle's curvature?

The perimeter of the square behaves like the circumference of the circle in that three apparently equal turns in a Euclidean plane, all in a clockwise direction, will bring one back to one's point of origin, where a fourth equal turn is made to arrive at the original orientation of the circuit. The four turns express the four directions of a Euclidean plane, which have previously been indicated.

They occur in the order: down to the first angle, left to the second, up to the third, and right to the fourth, which completes the circuit. In addition to this circularity, the vertex of each of the angular rotations of the square's perimeter is demarcated by a radius of the circumscribing circle, this radius being half a diagonal of the square.

The square is inscribed within the circle, and its four sides mark off equal arcs of the circumference, which together fully circumnavigate the circumference. So it is fair to say that the number of degrees traversed by the perimeter of the square is equivalent to that of the circle. It is 360°.

And, since the sides of the square are the same, the four successive turns departing from and arriving at a point of origin on the square's perimeter must be the same. Thus the exterior angles of the square's perimeter are each one quarter of 360°, or 90°. It is this complete rotation within the plane, in combination with an identity between diagonals and diameters, that relates the square to the circle.

Note that an equilateral triangle does not fulfill this role. This in spite of the fact that it too can be circumscribed within a circle. And its sides, acting as chords, divide the full circumference of the circle into equal arcs. The exception arises from the fact that a triangle's three angles do not represent the full revolution of a circle. Though they enclose an area, the three orientations of straight lines do not express the four directions of a Euclidean plane.

For this reason, the area they enclose does not equal the area of a quadrilateral. Thus the area of a triangle is only half that of a regular quadrilateral and some portion of an irregular quadrilateral. Its degree measure equals that of a semicircle. And it is the only polygon whose internal angles are in sum less than 360°.

Now very little imaginative effort would be required for variations of the right triangle to be derived from the right-angle isosceles triangles which have been discussed as equal components of a square. It is a matter of extending one of the equal legs of a right angle without changing the other. An angle with unequal legs would then be formed. And adding a new third line to this would result in a non-isosceles right triangle.

So, if two of these non-isosceles right triangles are congruent, they can, at the line subtending their right angles, be joined together in such a way as to form a quadrilateral with equal long sides and equal short sides. The line subtending both right angles would be the diagonal of the quadrilateral. But, when the figure is drawn, it will not be a square. For its long sides are not equal to its short sides.

The Limits of Reason

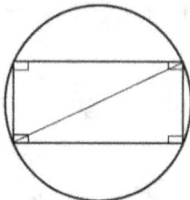

Nevertheless, like a square, the result can have vertices which appear to be tangent to the circumference of a circle which circumscribes them.

The fact that all four vertices of the quadrilateral tangentially touch the circumference of the circle, while the circle continues to look like a circle, illustrates the probable validity of the circumscription. So, consequent upon this circumscription is the fact that the shorter sides of the quadrilateral will mark off equal segments of the circle. This will be the case with the longer sides as well. For one can lay the short segments on top of each other and the long segments on top of each other to see that this is the case (or as near to the case as can be physically demonstrated).

This equality of segments in pairs results from the shorter sides of the quadrilateral being equal, as are the longer sides. The sides are the chords of segments of the circumscribing circle. Consequently, the segments are also equal in pairs. And each pair of adjacent unequal segments formed by unequal chords meets where these chords are tangent to the circumference and forms a vertex of the quadrilateral.

So the sides of the quadrilateral are four chords of four segments of the circle. And the four segments of the circle delineate arcs which together constitute the full circumference. The chords of the segments of the circle meet at the vertexes of the quadrilateral. Thus the perimeter of the quadrilateral appears to be proportionately linked to the circumference of the circle, as was the case with the circumscribed square.

But an inequality between adjacent sides of the quadrilateral exists. For that is the character of this quadrilateral. As its four sides are assumed to be successively perpendicular to one another—since they were so in each of the two congruent non-isosceles right triangles which were joined—all four vertices of the quadrilateral will be right angles.

Because the four vertices divide the full circumference of the circle into arcs, the perimeter of the quadrilateral would appear to be 360°, as was the case with the square. And the graphic illustration of this quadrilateral circumscribed by a circle cannot be satisfactorily drawn in any manner but that which appears to make the quadrilateral possess four right angles.

So the quadrilateral is a four right-angle figure, each angle enclosed within a pair of unequal sides. But that is the point: the pairs of sides enclosing the angles are not equal to each other. So they form a right-angle quadrilateral which is not a square. They form what shall be called a rectangle.

This rectangle can be inscribed within a circle, as has been made graphically evident. Moreover, its diagonals also function as diameters of the circumscribing circle. So it has a relationship to the circle which is somewhat like the square. However, its adjacent sides yield unequally-sided relationships with its diagonal, as in the case of the 3, 4, 5-sided Pythagorean right triangle and others like it.

For this reason, it does not exhibit as close an affinity to the circle as does the square, which forms an equal- and thus symmetrically-sided Pythagorean relationship with its diagonal. The circle is perfect symmetry. And the square exhibits a symmetry more closely resembling that of a circle than does the rectangle.

The Limits of Reason

It can now be seen that a square, a right angle, a right triangle, and a rectangle are all initially imaginatively conceived, only to subsequently become concepts by means of precise definitions. For conceptualization does not govern imagination. It follows it. Imagination remains a free exercise of the mind, limited only by the available impressions on the mind derived from physical experience. It works its magic through an unfettered association of properties. Whereas conceptualization only follows the work of imagination when definitions are set in place, binding the image to the requirements of the definition.

Accordingly, once images of perception are taken up in thought, they may be modified into ideal images much in the manner that the properties derived from the images of a horse and a narwhal may be combined and made into an image of a unicorn. For, while there are horses and narwhals in physical experience, there are no unicorns.

The imaginative images of geometry have initially been taken from experience in the same manner that the imaginative images of a horse and a narwhal are taken from experience. Thus a perfect circle is an imaginative idealization derived from the imperfect circles of experience. Likewise, the right angle and right-angle figures are so derived.

However, like those of the unicorn of free imagination, the properties of circles, right angles, and right-angle figures, which had their origin in physical experience, have subsequently been amended by free imagination into something not found in physical experience. For example, the circle is imagined to be *perfectly* round. Or a right angle is imagined to be perfectly perpendicular. So the new image becomes an idealization which is simply imagined.

But this image, left in its imaginatively flexible and therefore undetermined state, is insufficient for rigorous thought. It must become a fixed concept to be made useful for the logical precision of reason. So the ideal image—a free and openly modifiable imaginative association of properties—is logically bound into a closed image, which is a concept and which cannot be further modified. It cannot be modified because its properties have become a part of a definition which imparts an inflexible rigor to it.

It is at this stage, as an image which has lost its imaginative flexibility through the application of a definition, that the concept of a right angle takes its place in Euclidean geometry as a definition. And it is in like manner that right-angle figures are derived as propositions from the definition of a right angle. For these ideal images have been transformed into ideal concepts. Consequently, none of them has any longer a correlation in physical nature. Any immediate connection to experience has been removed.

What this implies is that geometry is not empirical truth. Nor is it innate to the mind. Rather, geometry, like any other imaginatively developed and logically organized system of thought, is no more than a template placed over the physical world. This is true of any geometry, not just Euclid's. All geometries, and all logical systems of any kind, are conceptual approximations of physical experience. They are each created by the human mind out of two elements: the individual mental impressions of physical experience and the operations of the mind.

So it is that the perfect circle, straight line, and right angle originate as conceptual generalizations. For these ideal geometrical concepts do not represent anything which exists in physical experience. Rather, they are derived from mental images which

The Limits of Reason

exist only in the imagination and ostensibly on paper. They are objects of the mind.

Now any object, mental or physical, is quantitative in the ratio of its parts. There is at least length, as in a line. And where there is a geometrical figure, as in the case of a circle, a polygon, or a polyhedron, there is length, breadth, and, in the latter case, depth. So it is a matter of deciding whether the properties in one figure are quantitatively equivalent to or different from the properties in another. For even the shape of a polygon or the face of a polyhedron is differentiated quantitatively in terms of the number and relative length of its sides. Thus the relations between geometrical figures are rendered comprehensible by means of a quantitative comparison.

Likewise, it is by means of an imaginative development similar to what has been the subject of this discussion, that the above figures may be transformed into others. That is, lines may be restructured into angles, circles into ellipses, squares into rhombi, etc. Moreover, insofar as they are geometrical concepts, it is from the seemingly more exact figures that the less exact are derived, not the other way around.

It is in this way that the concept of a circle has evolved into so many concepts of various ellipses and curves (the latter both closed and open), that the concepts of a straight line and a circle together have contributed to the angle, that the right angle and the perpendicular line were conceived in close support of one another.

Going beyond this—and this is an important issue in Euclidean geometry—is the fact that figures constructed with straight lines, like the polygon, must be brought into association with figures such as closed and open curves. This relationship is established by means of the various angles which serve as an imaginative and conceptual bridge between straight-line and curved figures. For it should be noted that the precise character of an angle cannot be delineated without a reference, directly or indirectly, to the circle or a curved potion of it. Yet angles are composed of straight lines.

Once an imaginative figure has been developed into a closed classification—such as a concept like the circle and its derivative semicircle, or a square and its derivative isosceles right triangles—it may then enter the logical discourse of mathematics. With the inevitable help of imagination, much more may be deduced from such concepts within the appropriate logical structure of mathematics.

That is to say, this may be done, provided no attempt is made to go outside the system of mathematical thought, where such concepts would then be examined on a strictly empirical basis. For in such a case it would be discovered that, due to their separate imaginative origins, uniform curves are incompatible with straight lines.

Even within the geometrical system, the incompatibility can result in an indeterminate measure. This is demonstrated by the uncomfortable relationship which arises from a conceptual union between the circumference of a perfect circle and its radius, which former is a uniform curve, and which latter is a straight line. For this yields the transcendental number pi.[17]

---

[17] This relationship has been discussed in the previous essay. Nevertheless, it should be noted that pi is not only an irrational, but a transcendental, number because of the indeterminacy of the relationship between the circumference and radius of a circle. Whereas the square root of two is irrational, but not transcendental, because the relationship between the sides and diagonal of a square is incommensurate rather than indeterminate.

The Limits of Reason

So let it be emphasized once again that what has been attempted in this discussion is an illustration of the fact that the associations of imagination both precede and underlie the operations of systematic reason and logic. It is imagination which is the creative force in human thought. Reason is its organizer. The former creates concepts. The latter systemizes them. But even the logical links of the system ultimately rest upon imaginative associations.

This is to say that the latter is of a concretely imagined character, whereas the former is not. The square root of two is an algebraic root which exhibits an incommensurate relationship between the sides and the diagonal of a square. In both nature and thought, what is incommensurate is more common than what is commensurate. This can be experienced and understood. But how many points on a circumference (i.e., how many radii) would be required to ensure the uniformity of the curve of the circumference of a perfect circle? This can neither be experienced nor understood.

George Lowell Tollefson

Logical Implication

Again let the circle and square be constructed. But let it be done in the simplest possible manner. What is being sought is the original inducement to develop these geometrical figures by means of the agency of physical experience. In other words, what would have suggested geometrical science in the first place?

A person standing on level terrain imagines a point above his head which is twelve feet from the ground.

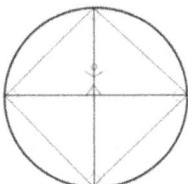

Marking the space between his feet with one end of a twelve foot rod, he allows its opposite end to fall to the right and left of him, describing a semicircle touching the ground on either side.

Thus these lateral points of contact with the ground are the same distance from the space between his feet as is the point above him. Now he imagines using the same rod to mark a point twelve feet directly below his feet. He has created a circle, the circumference of which is constructed with the rod as its radius.

He then imaginatively connects with a straight vertical line the previously indicated points of the circle's circumference above and below him, repeating the process to his right and left

The Limits of Reason

with a horizontal line. He connects the four ends of the two lines with straight lines. In doing so, he has created a circumscribed quadrilateral. In the drawing above, the sides of this quadrilateral appear to be equal. And the four vertices formed by them also appear to be so similar to one another as to suggest the possibility of their being equal.

The imaginative conjectures being proposed here are strengthened into something near conviction by the figure. So the visual appearance of that figure has strongly suggested a square and four trilateral polygons, which latter are considered to be congruent with one another because their sides and angles appear to be the same. This was previously discussed in a slightly different manner.

Most importantly, it has been accomplished by imagination, not by logical proof. It is empirical in the sense (and only in this sense) that it is composed of elements suggested by physical experience. But what has been constructed from these elements exists only in the mind. For the circumference of the circle is proposed to be a uniform curve.

Of course, one might arrange the situation so as to deduce proofs by logical implication. But proofs can only create a condition of unity in the mind, and thus a somewhat illusory sense of practical certainty. The proof is a unity of thought, not of perception. The sense of certainty in the application of the unity of thought to the perception would be no more than analogous in character.

The analogy arises from the fact that certain objects in the mind are assumed to be identical with objects in physical experience. But, in fact, the mental apprehension of a physical object varies slightly according to each projection of the image of the object on the mind. For the object exists in the mind as an imaginative unity of impressions of its individual qualities. And these differ somewhat each time the image is repeated in the mind. Some qualities are omitted, others added. So it is a generalized image which is formed into a more durable concept.

The concept exists within the mind. It is not external to the mind. Nor is it founded upon an exact replica of the object. So it can only be understood to closely suggest a parallel with the object. Consequently, any proof constructed with such concepts cannot create an incontrovertible conviction of reality.

Rather, a sense of certainty concerning the proof would be derived from the unity of the logical implication. The implication is an association of concepts by means of those properties which they either have in common or, if not, are supposed to have in common. The mind incorporates the concepts involved as associated terms within a statement. Since this occurs within the unity of consciousness, it creates a sense of unity in the statement. And a kindred sense of unity is likewise sustained between similarly associated statements. Hence the process of logical implication which results.

Thus logical inferences made in this way would appear to arise from their associated relationships. But an analogy between logical inference and physical experience is not a certainty. It is an act of will. For it is the will which has created the associations between the properties of the concepts involved. It is the will which has held the mind to those associations. And it is this same will which identifies the foregoing mental processes with physical experience.

So let the reader's attention return to the above circle and square. If a person imagines the two intersecting diagonals he has posited (for that is what they are) to be the one horizontal, and the other vertical, to the ground, and thus perpendicular to

The Limits of Reason

one another, they will create four apparently equal angles at their point of intersection.

Four right angles can be imagined, as was considered in the previous essays. And, as was also shown before, the construction of isosceles right triangles and other types of right-angle polygons can be derived from the resulting circle and square. Moreover, the angles within these figures can be modified to produce non-right-angle polygons. In all cases, proofs will come later. They will follow upon the original imaginative efforts.

Thus it can be seen that even the simplest form of imagination has a role to play in the construction of what are conceived to be the most precisely reasoned figures. For example, a wheel might have been rounded through rough experience. A circle suggested by the wheel. Or the noonday sun might have been observed in the sky and imagined as a point. The direct line of sight between that point and a person becomes a straight line. The straight line becomes a moving radius. The moving radius could describe a semicircle, much as the sun appears to move in an arc from horizon to horizon. Etc.

But what is most important is the role of imagination in logical relations. Because concepts are created from imaginative images, they share with them their condition of uncertainty in relation to experience. Some of the perceptual impressions on the mind which accompany any physical experience are inevitably omitted in the mental representation of that experience.[18]

---

[18] As this is being discussed from the representational viewpoint, an immaterial account of the same phenomenon would be somewhat different. Instead of maintaining that some mental impressions are omitted, it would assert that there is a repeated series of associated mental impressions which constitute the object. And, since the repeated associations of impressions

Even the most thoroughly representative images are simplified in the mind in this way.

So, of necessity, the representational image does not include all the impressions on the mind which constitute a perceptual experience. It is approximate. Hence concepts formed from such images are also approximate. Consequently, any association of these concepts in structures of logical implication, however rigorous the appearance of their mental relations, is suggestive of physical experience rather than certain.

And this fact is not changed by repeated experimental verifications of the mental relations. For the observed underlying causal relations are at best inductive in character and therefore no more certain in their consistent repetition than the thought process is certain in its fidelity to such experience.

---

are not completely identical, this accounts for a generalization and reduction of the impressions, or qualities, in the cumulative mental image of the object. This is the correct but less practical way of stating the issue.

Squaring the Circle

Archimedes' proof for his first proposition from the "Measurement of a Circle" is an illustration of the rational incompatibility of the circle with a linear figure. A close examination of this relationship suggests the ideal character of all geometrical concepts and, by implication, the whole of mathematics.

The proposition is as follows.

> The area of any circle is equal to a right-angled triangle in which one of the sides about the right angle is equal to the radius, and the other to the circumference, of the circle.[19]

Today this proposition can be demonstrated algebraically. Take the area of a triangle at 1/2bh and substitute radius r for the height and the circumference $2\pi r$ for the base. This substitution is what Archimedes did figuratively in constructing his equivalent triangle. It yields $A = 1/2(2\pi r)r$, which is $A = \pi r^2$, the area of a circle. But, not having this algebraic method to simplify matters, Archimedes chose to conduct his proof by means of a double reductio ad absurdum.

The reason for his doing this was that he could not square the circle because pi determines its area in such a manner as to render it indeterminate. That is, if he wanted to work only with de-

---

[19] Archimedes, "Measurement of a Circle," Proposition 1.

finitive rational measures, he could not do so because he could not then describe the base of the triangle he had in mind.

Since its height represented the rational radius of a circle, its base measure must include a transcendental irrational, which determines the circumference of that circle. Thus the area of the right triangle he conceived as an equivalent to the area of a circle must be indeterminate in the manner that the area of a circle is.

The triangle is indeterminate in area because it is not a linear figure rationally bounded at its vertices. It only appears to be so, since the base extending from its right angle is indeterminate. Thus the area of this triangle cannot be expressed in a rational measure for the same reason the area of a circle cannot be so expressed.

In other words, the area of a circle is expressed by $A = \pi r^2$. The same formula expresses the area of the triangle Archimedes constructed. Since the areas of both the circle and the triangle have the same formula, the same is true of both. This is because π in the formula represents a transcendental irrational, which number is a constant linking the radius of the circle with its circumference. Thus it renders at least one side of the equation numerically irrational. Which is to say that, for the radius to be rational and non-zero, the circumference would have to be pi or some multiple of pi.

Leaving Archimedes and proceeding further in pursuit of an exploration of the ideal, let a different trilateral figure be considered, such as an isosceles right triangle. As previously noted, an isosceles right triangle is definitionally ideal in the sense that its sides are incommensurate with its base. This is because its hypotenuse will be irrational if the legs extending from either side of the right angle are assumed to be identical and rational.

The Limits of Reason

In other words, the hypotenuse in this case is invariably one of the identical sides multiplied by the square root of two.

To illustrate, the Pythagorean formula for an isosceles right triangle is $c^2 = 2a^2$. So, if each side "a" is the square root of two, or a rational multiple of the square root of two, the hypotenuse "c" will be rational. That is, if the irrational number $\sqrt{2}$, or three times the irrational number $\sqrt{2}$, is substituted for side a, the result will be $c^2 = 2(\sqrt{2})^2$ or $c^2 = 2(3\sqrt{2})^2$ respectively. Thus the hypotenuse is either 2 or 6, both rational numbers, while each side is either $\sqrt{2}$, or $3\sqrt{2}$, both irrational numbers.

Conversely, if the sides are rational, the hypotenuse will be irrational. Thus, if each side is 2, then the hypotenuse is $2\sqrt{2}$. Or, if each side is 3, another rational number, then the hypotenuse is $3\sqrt{2}$, etc. Consequently, either the hypotenuse or the sides must be irrational. For the sake of brevity, the hypotenuse will be assumed to be the irrational measure.

So, if a right angle is constructed with two equal and rationally measured legs enclosing it, it appears that an isosceles triangle made with this angle could be described with a straightedge ruler. For it would seem that all which needs to be done is to join those legs with a straight line to get the irrational line which is the hypotenuse.

There is no need to define the measure of this adjoining straight line. However, it is clear that describing such a line independently would demand that the length of the line be measured. And this would necessitate creating a line of irrational measure. In other words, it would necessitate creating a line which cannot be created.

So drawing a hypotenuse by connecting the two legs of an isosceles right triangle by means of a straightedge does not mean the precise measure of that hypotenuse is known, any more than it can be asserted that a right angle has been created. In both cases, a measure is not taken. An assumption is made.

Now let this analysis be extended further. Let an equilateral triangle be considered. It can be seen that this example is more complicated because, by definition, the sides of the triangle cannot be incommensurate with one another. In addition, neither is a right angle being dealt with anywhere within the triangle. All three angles are what Euclid refers to as "acute."[20] They are "less than a right angle."[21] This can be visually verified.

It is assumed that in a Euclidean plane, when three equal sides are to be measured and put together as a trilateral polygon, a triangle of 180° is obtained with three interior angles of 60°. But this fact must be logically deduced. It cannot be demonstrated in a physical manner. For there is no means of an angle's measure being empirically observed or obtained with indubitable accuracy. If this were to be attempted, measures close to those anticipated could be obtained. But there could be no certainty in the result.

It does not take much of a further development of this line of reasoning to reveal that the same applies to all triangles. They are all idealizations, just as surely as the circle and straight line are. Assumptions must always be made before assertions incorporating these assumptions can be laid down.

This is not only true in Euclidean geometry. It extends to other geometries and ultimately the whole of mathematics. For it reflects the ideal character of all logical systems. Every concept, however close to its empirical source of derivation, is ideal

---

[20] *Euclid's Elements*, Book I, Definition 12.
[21] Ibid.

The Limits of Reason

to some extent. It is not identical to experience. No mental image is an exact reflection of a physical object.

In Euclidean geometry, as has been noted previously, it may be observed that an angle is a hybrid idealization lying conceptually between a circle and a straight line, which are themselves mutually exclusive and independent idealizations. Angles are thus an idealization built upon others. So the concepts prior to the angle (the circle and the straight line), being themselves idealizations, may be considered assumptions inasmuch as their role is ideal, or hypothetical, from an empirical standpoint.

In short, this ideal character of geometrical figures, allowing for an irrational connection between the circumference and straight line radius (or the straight line diameter) of a circle, is the reason Archimedes could not square the circle. The same problem of an irrational relationship, which exists between the radius and circumference of a circle, holds for the equivalent right triangle created by Archimedes. It also holds for an isosceles right triangle, though the irrational number $\sqrt{2}$ is algebraic and not transcendental. It holds for the isosceles right triangle because of that triangle's symmetrical relationship to the circle, which was discussed earlier in connection with the inscribed square.

This symmetrical association of the isosceles right triangle with the circle brings the isosceles right triangle's irrational, and consequently ideal, character to the fore. For its internal relations are like those of the circle. But a geometrical concept need not involve irrational relations to be ideal. It need only be imaginatively rarified and thus removed from the empirical realm of physical experience.

Accordingly, any triangle will exhibit an ideal character. If it were for no other reason, this would be the case because the measure of its legs is only supposed to be known. For to obtain such a measure the legs must be related to something outside the triangle, such as a straightedge rule.

Thus the measure of a triangle's legs is only purported to be known because no such measurement is exact. For this reason, and the fact that all numbers are themselves ideal, numbers are assumed for the measurement. But at least the use of a straightedge and numbers provides a feint in the direction of empirical representation.

The angles, however, are different when it comes to measurement. For the 180° measure of a triangle is not related to anything outside the conceptual realm of geometry. Rather, it is logically related to the 360° measure of a circle. But degrees are an arbitrary measure. They are a division of a whole within a plane.

Thus the circle is a whole round. Starting at point A, its circumference comes round to point A. And a triangle, equivalent in degrees to a semicircle, is a half round, though the perimeter of a triangle does complete a circuit of its area. But, if it were a whole round, it would be a quadrilateral, like a square or a rectangle, both of which must be composed of two triangles. These relations can be imaginatively seen. They can be determined by association, as in noting that two triangles make up a quadrilateral. But to be derived within a mathematical structure of thought, these relationships must be logically deduced.

Logical implication is itself ultimately associative. But the associations are not between images. They are between definitionally bound concepts. Which is to say that they are between the precisely delineated images of these concepts. Consequently, the process of logical implication demands an exact comparison of properties.

The Limits of Reason

It is for this reason that, other than to establish proportionate relationships between one thing and another, degrees express nothing about circles or angles which is independent of their intrinsic spatial character within a plane. For degrees are properties which are completely independent of empirical reference.

So, if one thing merely stands in some numerical ratio to another, and nothing else concerning either of them is being asserted, then, in determining that relationship alone, it does not matter what either of those things is independently of the character of that relationship. This is a clear example of the self-enclosed character of a geometrical system. For the ratio is definitionally determined within the system. It is also an example of the logically self-enclosed character of mathematical systems in general.

However, radians and the trigonometric relationships of the sides of a right triangle (and any other triangle by derivation) do provide something of a way out. They provide a quasi-connection to the physical world. Radian measure allows the introduction of the straightedge rule, which has purportedly measured the radius of a circle as a unit 1, rendering a circle's circumference in radii, or radians, as simply 2π. The 2π radians is equivalent to 360°.

Likewise, trigonometry relates measurements of the sides of a triangle in a similar way. But, once again, if the measure of angles is trigonometrically determined, what is discovered is a proportionate relationship (unfortunately, often incommensurate) between the triangle's sides. The only reason such a measure yields something practical is that the legs are purportedly measured with a straightedge rule, which is a part of the physical world.

A great deal in the physical world can then be more or less determined because the ideal (or conceptual) proportions of these geometrical figures can be related to the real (or perceived) proportions of physical experience. The relationship is, of course, approximate at best. For neither a geometrical figure nor an independent empirical measure is determined to an irrefutable exactness.

So it is important to remember that geometrical figures are entirely ideal. None of these figures can be constructed with absolute empirical accuracy. For no part of them can be measured precisely. Rather, when employed to solve practical problems, the figures constitute a template placed over physical experience. They are no more and no less arbitrary in their fit than the coordinate numbers on a military topographical map are to the actual terrain distances represented by the map.

On a topographical map there are contour lines. Each of these bears a relationship to the other contour lines on the map. That is, their relative concentric position and their distance apart determine the elevation and gradient of the terrain. This identifies land features like mountains, cliffs, and valleys. And it graphically displays relations of placement among symbols on the map.

The symbols may represent rivers, nonmoving bodies of water, etc. What is important is that they are merely symbols. They are lines and colored areas. So they disregard much of what constitutes the features which are being symbolized. Their purpose is simply to stand in for them in terms of extent and location: Here a river or lake is of such-and-such a length or size and at such-and-such a horizontal distance and vertical elevation from some other river or lake. Thus each feature covers a

The Limits of Reason

certain amount of specific terrain measured at a particular series of points.

The overall map in turn bears a closely approximate proportional relationship to the actual physical terrain. So a person identifies the location of a feature on the map in terms of two sets of number coordinates, which are determined by sequentially numbered vertical and horizontal grid lines on the map. These number coordinates together designate specific areas within a two-dimensional plane.

Now the relation of the numbers to the various features on the map must be kept in mind by the person using the map. In this way, she makes a comparison, matching several features on the map with the actual terrain she occupies. She estimates elevation, degree of inclination, and distance between the features to see if the map resembles the landscape.

Having done this, she can then establish her own position within the terrain, by identifying the land feature she is occupying and using the coordinate numbers on the map to locate it. This makes it possible for her to know where she is on the ground and to have some idea of where other things are in relation to her location.

Thus, like the degrees of an angle in relation to a circle, these ideal coordinate number relations, contour lines, and symbols on a map are irrelevant to anything beyond the map, until they are translated onto an equivalent terrain by means of a knowledge of how they are roughly proportional to it.

So it can be seen that it is proportions which govern the process of making the comparison. It is proportions which relate both the topographical map and geometry to the physical world. But neither is a contour or feature on a topographical map nor a geometrical figure itself an element of the physical world. It is simply a symbol, an idealization.

Euclid's Triangle

A triangle, like a perfect circle and a line without breadth, does not exist in nature. As has been previously stated, it is an idealization. Take the Euclidean line without breadth. When it is drawn, it is a quadrilateral having four sides, however narrow its width. Should a person imagine tapering it at one end to bring two sides to a point, it cannot be done.

This is because a vertex formed by two lines must have some width, which is at least the width of the drawn lines. So the vertex is "thick" at the joining of the lines, forming an overlooked *side* of the figure at the point of that joining. Otherwise, the vertex could neither be seen, drawn, nor detected with a microscope.

Thus, where two lines come together, there must be at least the width of one of those lines and a portion of the other at the vertex. So on paper there are no triangles, since no drawing of one can be made which correlates precisely with the concept. Nor can one be found anywhere in physical experience. Triangles are idealizations, albeit very useful ones, as ancient Egyptian farmers well knew.

But the question remains: how are they useful? The answer is: they are useful in terms of the proportions they exhibit. The human mind interacts with physical experience to create within itself a pattern of that experience which is conducive to proportional arrangement. Then it invents proportional geometrical relations so it can make use of them in developing a template to organize the physical.

The initial pattern of physical experience is determined outside of human control. For the mind's apprehension of that experience follows the sequence in which individual impressions are introduced to the mind. There is a fixity in this, lying beyond the human will. It is a predetermination in physical awareness which both assists and restricts the mind in its mapping of experience.

So it is here, where one thing follows another sequentially, that human experience is found to be outside of mental control. But it is also here that predictions can be made. In accomplishing this, mental constructs like geometry organize physical experience by matching a system of proportions to the general order of that experience. As a result, great facility in making measurements and predictions can be achieved.

The 5th Postulate

The difficulty raised by the various attempts to prove Euclid's 5th postulate may be philosophically resolved. For it is a conceptual problem, which occurs in Book I of *Euclid's Elements*. There is an apparent disjunction between two concepts which do not appear to be logically compatible.

Geometry is a science of logically organized spatial relations. This is what is referred to in the postulate.

> That, if a straight line falling on two straight lines make the interior angles on the same side less than two right angles, the two straight lines, *if produced indefinitely*, meet on that side on which are the angles less than the two right angles. [italics added][22]

Here the concept "produced indefinitely" is stated. But the lines are presumed to meet. Consequently, the emphasis is not on the indefiniteness of length of the two straight lines, but on the relationship of angles governing the mutual orientation of the lines.

It is stated in the postulate that the interior angles on the same side of the intersecting line, being less than two right angles, make the two intersected straight lines meet. Thus there

---

[22] *Euclid's Elements*, Book I, Postulate 5.

must be a reason why they would not meet, such as that the two interior angles are not less than two right angles.

So, if the same two interior angles were not less than two right angles but were equivalent to them, it is implied that the two intersected straight lines would be parallel, as stated in the 23rd definition.

> Parallel straight lines are straight lines which, being *in the same plane* and being *produced indefinitely in both directions, do not meet one another in either direction.* [italics added][23]

But to state that two straight lines are "produced indefinitely in both directions," that they are "in the same plane," that they "do not meet one another in either direction," and to add nothing more is to place an emphasis on the fact that these line cannot meet.

Now a line must be located within a plane in order to be a line. The definition of a straight line makes this clear.

> A straight line is a line which lies evenly with the points on itself.[24]

This is to say that the straight line lies on points which are in a Euclidean plane. So the line cannot be said to exist other than within the plane.

Beyond the assertions of indeterminacy of length, that the lines are straight, and that they do not meet, coupled with the

---

[23] *Euclid's Elements*, Book I, Definition 23.
[24] *Euclid's Elements*, Book I, Definition 4.

The Limits of Reason

fact that a straight line "lies evenly with the points on itself," nothing else has been given to condition the character of the two lines in the 23rd definition. To state that they do not meet along their indeterminate lengths is to add no useful information in addition to their indeterminate lengths. For, since those lengths are in fact indeterminate, it cannot be determined what it is to say that they either would or would not meet at some indeterminate point.

Geometry, in its original Greek form, is a science of comparative measure without number. And it is the 5th postulate which expresses concrete spatial relations, that, it would seem, ought to have been derived and not stated as an axiom. An example illustrating these spatial relations in some detail is appropriate.

Let two straight lines A and B be intersected by a third C at separate points D and E. Let it further be the case that lines A and B produce in relation to line C the interior angles D and E, which are together less than two right angles. This is the 5th postulate, which is expressed in accordance with the geometry of a Euclidean plane.

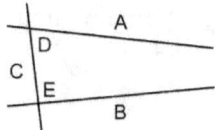

Thus the postulate states that, when lines A and B are extended from the intersecting line C on the side of the interior angles, they will meet at some point.[25]

---

[25] *Euclid's Elements*, Book I, Postulate 5.

George Lowell Tollefson

To demonstrate the source of the problem encountered by attempting to discover proofs for this postulate, attention will be focused on a circle. Let a circle be described about a center point A. From this point let two radii be extended to the circumference. At their respective points of intersection with the circumference at B and C, let a third straight line be drawn between them.

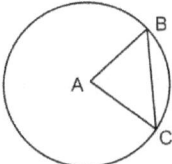

The triangle thus formed corresponds to the suggested triangle in Euclid's postulate. The acute interior angles formed at B and C are angles D and E of the postulate, as represented in the previous figure.

ABC is an isosceles triangle, two sides of which are radii of a circle. And the third side is a chord of that circle. As the length of the radii is fixed, only the length of the chord can change. But there is a limit to how much the two base angles within the triangle can be decreased, since they must define an area within the triangle.

Angles B and C become more acute as line BC is lengthened. Thus, as the third angle enlarges to approach 180°, the two base angles decrease accordingly. This occurs until line BC becomes straight line BAC, which is twice the radius, or equivalent in length and location to a diameter of the circle.

Thus on either side it marks off a sector which is a semicircle, and which is 180° and exhibits no angles. The two sectors together are the 360° of a complete circle.

The Limits of Reason

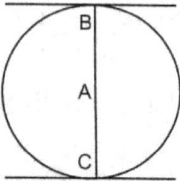

So at points B and C, let tangents to the circumference of the circle be drawn. They form right angles with line BAC. Thus the right angles at B and C, on the side of line BAC that triangle BAC was on, are together equivalent to the three collapsed angles of that triangle.

Now, returning to the original triangle, let angles B and C become less acute, each approaching 90°, as line BC of the triangle approaches 0 and lines BA and AC merge into a single radius. In other words, as the chord BC is shortened toward 0, lines BA and AC approach one another until they become one line. Once again, the triangle is collapsed. And what remains is a full circle.

Now let a tangent to the circumference of the circle be drawn at the point where the two radii merge into one line and line BC becomes 0.

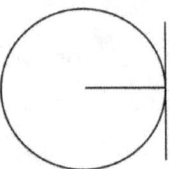

The intersection of the tangent with the line representing the two merged radii produces two right angles. These again represent the collapsed triangle, or 180°. Thus it can be seen that two interior right angles could not have been formed at vertices B and C within the previously existing triangle ABC prior to its collapse.

So a triangle is bounded by a semicircle and by its own non-existence. Between these extremes, it continues to reflect the 180° character of a semicircle in the mutual orientation of its angles. In other words, at any point in the expansion of the base of the triangle, its three angles, if lined up adjacent to one another, would describe a straight line which, when composed of two radii at the full expansion of that base, would demark adjacent semicircles.

In regard to these observations, it should be recalled that the concept of angles is derived from the concept of a circle and thus bears a close conceptual relationship to it. So the fact that a triangle corresponds in the sum of the degrees of its three interior angles to the 180° measure of a semicircle needed to be emphasized.

Consequently, if, as in the 5th postulate, two acute angles are formed by three intersecting straight lines, where the two lines extending together from the two acute angles approach one another, the two lines will meet and realize the form of a 180° triangle, unless a fourth line is added. It is as a consequence of this insight (made without the concept of degrees) that Euclid's 5th postulate was originally proposed.

But what appears to be inexplicable is the 23rd definition. How can it be known that two straight lines extended indefinitely will not meet? The answer must lie in Euclid's combined definition of a right angle and a perpendicular meeting of lines.

> When a straight line set up on a straight line
> makes the adjacent angles equal to one another,
> each of the equal angles is right, and the straight

The Limits of Reason

line standing on the other is called a perpendicular to that on which it stands.[26]

This is nothing other than the imaginative observation that, if a straight line should intersect a second straight line in such a manner that it is no closer to the second line on one side than it is on the other, it is perpendicular to that line.

What this implies is that a third straight line will be parallel to the first if it intersects the second line in the same manner that the first does. In other words, if it exhibits the same characteristic of perpendicularity as the first, but is not in the same location on the second line (to which they are both perpendicular), it will not meet the first line. For, to do so, it must approach nearer to the second line on one side or the other. All this assumes a Euclidean plane. And it is, of course, an imaginative observation. But it is not by any means illogical.

Thus it can be seen that both the postulate and the definition arose from imaginative origins. Euclidian concepts were then formed from these images. It is these concepts which are the 23rd definition and 5th postulate in the first book of the *Elements*. Any logical implication which might follow from a comparison of these two concepts would then arise as the result of an association of their stated properties.

For the conditions of the 23rd definition are established in such a manner that they would substantiate the conditions set forth in the 5th postulate: one straight line inclined, however slightly, toward another would intersect it at some point. This would occur in accordance with the character of angles, as set

---

[26] *Euclid's Elements*, Book I, Definition 10.

forth in definitions 10, 11, and 12 of Book I of the *Elements*, and as observed in their relationship to a circle.

The Limits of Reason

Parallel Lines

Let a mechanical illustration concerning parallel lines be conceived. Begin by replacing the two interior angles of Euclid's 5th postulate, which are less than two 90° angles, with two interior angles which together are equivalent to two 90° angles. It is assumed that they are not perpendicular to their intersecting line. So the angles are not the same. Thus there are the three straight lines: two parallel and another which transects them both non-vertically.

Now let it be imagined that these three lines are metal bars, the transecting bar and the two parallel ones being of unspecified finite lengths. The two parallel bars are fastened to the transecting bar at a specified finite distance from one another. And at its center the transecting bar is fastened to a supporting base in such a manner that, when the transecting bar pivots on its base, the two parallel bars will pivot at the two points of intersection with it, moving them horizontally and simultaneously displacing them vertically without altering their parallel orientation.

Initially, each of the parallel bars is transected at supplementary angles of 120° and 60°: 120° for the interior right angle of the top bar and 60° for the interior right angle of the bottom bar.

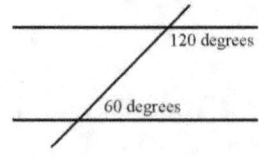

Thus the interior angles of the two transected parallel bars to the right of the transecting bar are together the equivalent of two right angles.

Now allow the two parallel bars to be pulled in opposite directions: the top to the right, the bottom to the left, while maintaining their parallel orientation, so that the interior angles remain supplementary at 180°.

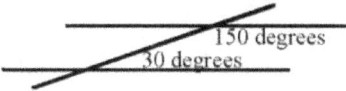

This results in several changes. First, as the two parallel bars move horizontally in opposite directions, they move closer together vertically. Second, the obtuse angle becomes more obtuse as the acute angle becomes more acute.

Say they have been repositioned at 150° and 30° respectively. Should the parallel bars continue to be pulled in the same manner, the vertical space between them collapses further. This condition progresses until the vertical space between the two bars is 0. And they are aligned. In other words, they lie upon the same line.

What does this simple mechanical operation illustrate? It demonstrates that, if the space between the two parallel bars is vertically collapsed, the parallel bars are seen to lie along a single straight line. When two straight lines lie continuous with one another as a single straight line, there is no longer any means of distinguishing their separate identity in a Euclidean plane. Thus parallelism is, in fact, single. For location in the Euclidean plane is the only distinction between parallel lines. Without this differentiation in location, they are the same line.

The Limits of Reason

This is an insight which leads one to believe that location was not an important consideration in Euclid's presentation of the 23rd definition and the subsequently conceived 5th postulate[27]. He simply had the idea that, if a transected line is defined in a certain way in terms of its angles and another line is defined in exactly the same way—that is, if the adjacent, interior, and exterior pairs of angles on one or the other side of the transecting line express supplementary relationships, and if adjacent angles across the transecting line are also supplementary—then the parallel lines are in practical terms the same line, regardless of the fact that they may appear to be separately located in space.

The appearance of separateness in space is simply a means of establishing the concept of a parallel relationship—a concept which is no less ideal (or no less initially imagined) than that of the perfect circle or the breadthless straight line. The space between parallel lines is, in fact, a band of consistent width. It is the equivalent in all respects but that of its width to a single straight line which is breadthless. For the parallel lines behave as if they were together a single line of uniform width.

---

[27] *Euclid's Elements*, Book I.

George Lowell Tollefson

The Angle's Origin

The problem with the Euclidean ideal concepts of the perfect circle and the straight line is that there is no logical relationship between them. Such a relationship must be systematically constructed in a proposition. However, the circle's radius is, in fact, a straight line. Though the perfect circle and the straight line are independent idealizations, the circle is a classification encompassing a straight line as one of its properties. Its properties are not logically combined as in a proposition. So the circle is simply a product of imagination.

It is an image created by imagination and not by reason—i.e., not by logical combination. As an image, it is composed of free associations of individual mental impressions originally garnered from physical experience. Many of these original impressions are filtered out in the creation of the ideal image.

Since it cannot be demonstrated that this image corresponds precisely to anything in the physical world, it does not express a direct reference to that world. It is a free image (or set of images) which is converted into a concept by means of the regulation of its properties by a definition. Thus it becomes an ideal concept. As such, it can be used in rational thought like any other concept.

However, the character of its origin does not depend upon the geometrical system in which is employed. Rather, the system assumes it as a building block for its subsequent demonstrations. In other words, perfect circles are incorporated into Euclid's system as an arbitrary act—an act of imagination and not reason.

The Limits of Reason

But angles are also included among Euclid's preliminary definitions. And the observation that angles are idealizations which attempt to bridge the gap between circles and straight line figures can be offered. But where is this laid down as a principle? There is no mention of this gap-bridging role.

Nevertheless, it would certainly appear that the concept of angles appropriates something from both circles and lines. For one of the two lines which bound an angle may be moved apart from the other in a counterclockwise direction in the Cartesian manner to form an ever more obtuse angle approaching and terminating in a full circle.

Or it may be moved toward the other in a clockwise direction to form a more acute angle approaching and terminating at 0. Yet the angle is composed of straight lines. And there is nothing present in either the concept of an angle or the figurative representation of it but these two lines and the distance between their open ends.

It is also true that the straight line radii of circles may be employed in such a way as to relate them by measure. This is demonstrated by Euclid in his first proposition.[28] Thus a means has been found in which an equilateral triangle can be constructed. And, since the sides of the triangle are of equal length, it is evident to imagination alone by means of a graphic representation that they exhibit a figure with three equal spacings between its adjoining lines. For a lengthening or shortening of any one of the sides would visibly upset the equivalency of the angles. But the fact that these angles express a relationship derived from the radii of circles is not mentioned.

---

[28] *Euclid's Elements*, Book I, Proposition 1.

Angles are a concept which arises from the properties of a circle, one of which is its radius. In the first proposition of Euclid's *Elements*, angles are introduced by means of the radii of two circles.[29] However, this critical relationship is not stated. For the proposition is concerned with the sides of a triangle.

But, had it not been for two principles, the equilateral triangle could not have been derived in the first proposition. One of those principles arises from the definition of a perfect circle. The definition states that a perfect circle contains a straight line radius which is expressed as a constant measure within that circle.[30] That means the concept of a straight line is incorporated as a property within the concept of a perfect circle and may be related to other radii. For all radii express the same fundamental characteristics.

The second principle involves the general concept of an angle. For it is indeed given without statement. Though the angle is not explicitly described as incorporating the concepts of circles and lines within itself, this is what it does. For it is a circular, or rotational, relationship of lines, as demonstrated by the relative measures of angles, particularly as formed by the radii of a specific circle. Understanding this, one can look inside the circle to find a common reference for the different relationships between two straight lines meeting at a point. That reference is the center of the circle. The relationships between radii projected from this center are angles.

---

[29] Ibid.
[30] *Euclid's Elements*, Book I, Definition 15.

The Limits of Reason

The Number System

A whole number, such as 24 or 25, is a multiple of the fundamental arithmetical concept, the unit 1. And, like the unit 1, it is an idealization. This can be demonstrated by considering the indeterminate character of such a number. It may be further demonstrated by observing the indeterminate relationship between any two such numbers which are positioned consecutively on the number line.

In mathematics a number is a concept. If any concept is found to be indeterminate in character, that means it does not have a clear relationship to a physical object. Nor does it have an innate origin in the mind. Furthermore, if two consecutive numbers are indeterminate in their relationship to one another, then the number line itself must be drawn into question.

At first sight, 24 and 25 may be considered to be discrete numbers. That is, it can be observed that the number 24 and the number 25, which immediately follows 24, are each multiples of the unit 1. These multiples differ from one another by precisely one unit. Since a unit is the limit by which whole numbers may be expressed, in whole number terms 24 and 25 are discrete numbers. They are discontinuous in individual character and discontinuous in relation to one another. In other words, the whole number 24 is neither more nor less than 24 units. And it is separated from 23 and 25 by one unit.

However, according to what is not an uncommon practice in arithmetical operations, these numbers may also be broken down into continuous increments. The number 24.999..., for example, exhibits this characteristic. For all practical intents, it is the number 25. But, by its being expressed in continuous increments to the right of a decimal point, it is no longer considered discontinuous in individual character.

In reference to the manner in which 24 and 25 are defined as whole numbers, there can be no question about their discontinuity, both in their individual character and in their relationship to one another. This is because these numbers are distinctly demarcated by means of units, both in regard to themselves and in regard to their relationship to one another. They are clearly conceived to be discrete numbers. They are therefore determinate numbers.

But, in arithmetical operations, it is supposed that the number 24.999... may be substituted for the number 25. The number 24.999... is thus considered to be the continuously incremental version of the discrete number 25. Since there is no end to the 9s which appear to the right of the decimal of this continuously incremental number, it is therefore indeterminate. But the number 25 is considered to be determinate. Why the contradiction?

More importantly, since it is not known where the 9s will end, it is not known if 24.999... will ever reach 25. In fact, though the distance between the two numbers grows ever more minute, there is certainly always some supposed distance between them. So 24.999... is less than 25. It is in effect not 25. Yet it is allowed to be so.

The number 24 and some other number less than one unit—i.e., a decimal followed by an indeterminately long list of 9s—are together assumed to constitute the whole number 25. In other words, no matter what 9 is chosen to momentarily terminate the progression, say 24.999999999999999999999...9,

The Limits of Reason

there is another which follows it, further closing the gap between the progression and 25.

Now clearly the number 24.999... is not the number 24. For it is almost a whole number beyond (or unit more) than 24. But neither does it quite reach 25. Yet, because it approaches ever so close to 25, until the distance between the two numbers becomes negligible, 24.999... is permitted to stand in for 25.

But this does not mean it ever reaches that number. So the numbers 24.999... and 25 each retain a certain discontinuity in relation to one another. Where they come together and one is allowed to be substituted for the other, they must nevertheless both in a sense be considered discontinuous in their relationship to one another. For neither number is the other.

However, in comparison to the whole numbers 24 and 25, which are held to be demonstrably discrete in individual character, the number 24.999... is not a discrete number. For its discrete character cannot be numerically demonstrated. So, in that sense, it is continuous. And, when brought into a direct relationship with it, neither is 25 a discrete number.

Thus, since 24.999... is continuous and consequently indeterminate, it is not discrete in the manner of an independent whole number. So, while it is maintained that 25, insofar as it is a whole number, is discrete and therefore determinate, it is said the number 24.999... is continuous and therefore not determinate. If they are to be imagined as being in the same number line, it must be confessed that they are quite different in character.

But let a number line strictly made up of whole numbers be considered. That would be a number line where the discrete numbers 24 and 25 are found. Whole numbers like 24 and 25 may be considered descriptive of this number line, since like them all the numbers of this number line are supposed to be discrete in individual character.

So, going a step further, can it not be asked: Are not even the 9s in 24.999... in some sense discrete numbers borrowed from such a number line? Are not all numbers represented by combinations of the digits 0 through 9, which are whole numbers? If it is answered that this is a matter of convenience in using symbols, it should nevertheless be pointed out that this implies an original bias toward whole numbers, which bias is then built upon in—dare it be said—contradictory ways by means of unlimited progressions behind a decimal.

The point is this: Why assert that 25 or any other whole number is a discrete number? How can it even be asserted that the unit 1 is a discrete entity? Why designate them as discrete, when in the same logical system continuous numbers are substituted for them? Perhaps, when looked at in its entirety, the system called arithmetic readily admits that the whole number 25 is not truly discrete.

It could be that, as a concept, the true character of the limits of the number 25 lies hidden in what is not overtly expressed in that concept. If 24.999... can be allowed to represent 25, it is a clear implication that such a whole number as 24 or 25 does in fact bleed off in one direction or another when it is allowed to do so. That is, just as 25 bleeds off in the directions of 24 and 26, any whole number bleeds off in the directions of its two nearest neighbors. At least, it does so when, for computational purposes, it is found to be convenient that it should do so.

For instance, when the continuous incremental expression 24.999... was used, it started at 24 and approached 25 by means of addition. That is, 24 approached 25 by means of proportionately diminishing increments: $9/10 \rightarrow 9/100 \rightarrow 9/1000$. The

The Limits of Reason

first of these diminishing increments of 9 was placed in the first position to the right of a decimal. Then the next smaller one was put in the next position to the right of that digit, etc., always proportionately reducing the magnitude of each progressive 9 in the manner shown.

So 25 could approach 24 by reversing the process, using the same digits as increments of subtraction. This would mean that the whole numbers are not truly discrete in any way. They are not only individually continuous, but continuous with one another. For each number increases or diminishes into an indeterminate unknown, the precise character of which may not be reckoned by calculation. Overall, taking all numbers into account, that "unknown" would appear to be a continuous unity, there being no definitive break between numbers. In other words, it is a unity the character of whose parts cannot be definitively determined.

By this, it can be seen that the concept of a whole number is an arbitrary convention. Such a conclusion is supported by the fact that any number can be observed to flow endlessly in the direction of continuity with another greater or lesser number. So discreteness in a number must be established by fiat. But, since a number's limits can only be sought while never being quite attained, it would seem that a number's precise character is an elusive truth. It is elusive because the human mind cannot precisely determine a number's limits.

But neither can they be denied. For anything which is supposed to have no limits whatsoever must be considered indistinguishable from anything else and therefore individually nonexistent. The human mind must therefore assume that a number has limits, be they discrete or be they continuous and beyond calculation.

If the limits are beyond calculation, they are nevertheless assumed to be expressed in a manner similar to discreteness. For each of the individual increments, taken to whatever countable extent, is finite. So it turns out that discreteness is a convention brought into being for the sake of convenience in conceptual operations like those of arithmetic.

By means of this same convention—i.e., creation by fiat—the first ten members of the number line of whole numbers, 0 through 9, have become the symbolic and conceptual components of all numbers. Thus, considered in themselves independently of their present function, the individual digits in 24.999... are each a whole number.

That is to say, they are the symbols and concepts which are used throughout arithmetic, complicating and redefining their roles as more complex operations in which to employ them are derived. Thus the entire representational system of numbers is derived from them. And the whole of arithmetic is based on them.

Of course, this means the entire system is a convention. It is like an egg which has many components. But the components must be arranged in a specific manner to become an egg. Change the arrangement. Put the albumen in the shell and the matrix of the shell in the yolk. And there is no egg. That means there are not only a number of components which go into the composition of an egg. There are also rules which determine how they are related.

So, because the number system is a convention throughout, the placement of any number in the number line is determined by rules which must be rigidly adhered to. Ignore or discard one of these rules, and the number system collapses. Its constituent elements dissolve. It is no longer a number system. And its numbers are not numbers. They are not numbers because they are not defined by a number system.

So, in effect, they dissolve into air. The individual numbers become nothing more than a collection of disassociated imaginative images. And these images are conceptually none too clear. For they are not clearly defined. Thus it is seen that the logical medium in which arithmetical relationships are established must be faithfully adhered to.

But, if this is so, can any of the numbers within that medium be considered to have independent existence as a concept? Can any number stand on its own without reference to other numbers and to the logical rules which maintain not only their relationship to one another, but their very identity, which is, in fact, that logical relationship? Could it be that such a number is simply a product of the human mind's persistence in maintaining it within the logical medium of the number system?

This system within which numbers are maintained as multiples of arithmetical units, this persistent medium, this "logic," is nothing more than a product of quantitative proportion and will. For, other than by means of a general and nonspecific reference to those potentials for proportion found in physical experience, arithmetic does not come from that experience.

Nor can anything other than a general concept of limited unity be understood as approaching an innate contribution of the mind. It is this concept which is the origin of the arithmetical unit. And the increasingly complex ramifications of the arithmetical unit are the logical system of arithmetic. The system is a collection of associated classifications, which classifications are aggregates of units held together by various rules. The rules are governed by the general principle of proportion.

Any number's role within this structure is initially imaginative. When a number is conceived alone as an aggregate of units, but without reference to other numbers, it is simply an association of units. Its conceptual reality comes into being with the number system. And this number system is maintained by force of will.

It seems as though logic were a power residing in an eternal realm of its own, expressing necessary relations. But, in truth, human choice, expressed under the constraints of proportion, is what determines the particular character of each classification within the arithmetical system. The relationship of one classification to another is made possible by associations between the contents of each classification.

For example, two units in one classification (the number 2) may be cumulatively associated with three units in another classification (the number 3). And these together constitute a classification of five units (the number 5). From this association of two classifications together (the numbers 2 and 3) and the inclusion of them within a third classification (the number 5), the operation $2 + 3 = 5$ is determined.

The only necessity in this procedure lies in the limits which have been applied to the two original classifications. The associations of units are limited to two and three units respectively. Then they are put together in a countable sequence. The new classification formed is labeled according to the final digit in its number sequence. That digit is 5.

The initial counting of units in each number is done according to a previously accepted number line. Then all units are counted together. In this way, 1 2 3 4 5 is obtained, which is labeled the number 5, since 5 is the last digit in the sequence. This is the operation known as addition. So it can be seen how

The Limits of Reason

utterly rudimentary the foundational components of arithmetic are. As previously mentioned, only two things in the number system have any correlation with either physical experience or a supposed innate faculty of the human mind.

The first is arithmetical proportion, which arises by suggestion from physical experience. This is due to the fact that any orderly system configured within the human mind, such as its "map" of physical experience, must be amenable to proportion. Material experience could not be understood, if proportion were not to be found potentially within it. So the mind seeks to arrange it accordingly. But not without paying due regard to the order of the sequences of physical impressions received within the mind.

Neither could numbers be understood, if proportional relations were not created both within and among them. Even in a contrasting of physical qualities, as between apples and oranges, proportion, however inexact, is implied in the comparison of so much red to so much orange, so much sweetness to so much tartness, etc.

The second structural element of the number system is the recognition of simple unity, which arises from a fundamental capacity of the mind to isolate and compare experience. Simple unity is imposed upon experience in terms of limits. It is in this way that the arithmetical system adopts its fundamental component, the unit 1.

The unit 1 is a limited entity without any properties other than that of being a unit and possessing a variable magnitude. This unit is ramified into numerous interdependent numerical classifications by means of proportion. Such is the foundational method employed to create the arithmetical system.

Now the proportions found in numbers are only in the broadest manner reflective of physical experience. For it is not the specific relations of physical extensions that determine the numerical system. Rather, it is proportion which is refined within that system. These proportions could have been arranged differently in arithmetic, say in a seventeen-unit-based system rather than a ten-unit-based system. But the arrangement would still be proportional.

The concepts of the arithmetical system share units. Thus the fundamental concept is the unit 1. As mentioned above, that unit is derived from a recognition of a limited unity. A generalized limited unity is what becomes the unit 1. More complex mathematical concepts, like the quadratic formula $x = \dfrac{-b \pm \sqrt{b^2 - 4ac}}{2a}$, involve compounds of arithmetical units held together by rules of proportion. This is the case with arithmetical numbers and the algebraic variables which stand in for them.

Each such compounding is both a single concept and a set of multiple concepts. Thus the quadratic formula is a single concept which is a set of various numerical concepts brought into proportional relations. It is accordingly both a single classification and a compounding of multiple classifications. The simplest and most fundamental of the arithmetic sets of classifications is the whole number line.

Thus it can be seen that numbers and their system are ideal. Specific concepts, like the whole numbers 24 and 25, are abstract and immaterial. They neither exist in the physical world nor in an innate realm of thought. The logic which links them in various arithmetical operations has neither a physical existence nor a transcendent existence in the mind. Both the numbers and the logic which defines them and binds them together are creat-

The Limits of Reason

ed. They are inventions of the thinking mind. They are there because the mind chooses to put them there.

George Lowell Tollefson

Numbers

Individual mental impressions are the foundation of material experience. But they are not initially presented to the mind as distinct in themselves, nor as differentiated into objects. Rather, they appear to human awareness as continuous, vague, and crudely differentiated. This is because the recognition of specific mental impressions and the objects they compose is a function of experience. It occurs when the mind interacts with a great number of these impressions and is impelled to differentiate and group them.

All objects of experience are represented to the mind by images, which are associations of mental impressions. But, as the order of presentation of these impressions in the mind is independent of human volition, physical experience is generally understood as arising through the senses. As such, it is encountered as an extended but otherwise unclearly differentiated revelation. Its recognition is further complicated by the fact that it undergoes occurrence, alteration, and disappearance.

The physical world remains objectively undifferentiated prior to the interaction of thought with it. Thus it is indeterminate in character. It may be compared to an irrational number, which is also indeterminate—i.e., lacking in clearly demarked limits. The clarification of physical experience into objects inaugurates a recognition of discontinuity in the mind. The discontinuous, being discrete in character, may be compared to a rational number, which is also discrete, or clearly demarked.

From this it may be seen that irrational (indeterminate) numbers correspond to undifferentiated awareness. Though they are

The Limits of Reason

products of thought, they resemble that which is made present to awareness prior to any interaction of the human mind. Only, in their case, there has already been an interaction with discrete numbers. And it is the operations performed on these which have generated the indiscrete numbers that are irrational.

But rational numbers are ideal. And nothing directly correlative to them exists within physical experience. They appear only in the mind as discontinuous intellectual classifications. They are a template which the human intellect throws over experience in order to understand it, or at least to make logical sense of it.

A example of this can be found in a pile of stones. Initially the mind experiences the pile, which has indistinctly variegated features. Then it recognizes the stones as independent and multiple and not parts of a single object. In addition, it invents the counting numbers, which are systematically discrete. It uses the counting numbers to bring order to the pile of stones. That order is number. There are, of course, other ways in which the stones are examined and organized, such as by texture, hardness, color, etc.

The fact that from earliest life human beings must, however crudely, conceptualize the world in order to meaningfully perceive and interact with it, does not alter the fundamental distinction between the operations of the mind and the data of what it begins to recognize as the senses. The latter is physical experience, which, as initially made present to human consciousness, is conceptually, even spatially, meaningless.

Thus the need for an involvement of cognition at the earliest stages of human awareness. The cognitive mind provides relational and functional *meaning*—i.e., objective distinction, spatial depth, and enumerative proportion—to the data of physical experience. So it is through cognition and its interaction with experience that human beings are able to relate themselves spatially, practically, even emotionally and thus morally, to the world.

That both the immediate data of experience and the conceptual classifications of the human mind are involved in early awareness does not belie the differentiation of thought from what is generally understood as sense perception. It simply indicates that human beings have a relationship with experience which is complex, and which involves from the beginning an intellectual awareness and interaction.

It is in this manner that the human mind constructs a system of numbers which is ideal, and which appears as a correlative to discrete phenomena. These numbers are also discrete, like the articulated phenomena of experience. Their indiscrete counterparts are the irrational numbers. These are made to function within the system of rational numbers. For they occupy positions filling the gaps between them.

Rational numbers are ideal in the sense in which the term "ideal" has previously been discussed in this work: They are an independent creation of the mind and do not directly reflect anything in physical experience. Yet, like articulated physical experience, they are discontinuous and discrete, which is the character of every fully realized imaginative or conceptual representation.

Conversely, irrational numbers, which are continuous and indiscrete, are, at best, partially realized concepts. They have a quantitative character, like pi, which is a specific but indeterminate number. It is indeterminate because it is incomplete, which is to say that it is not rounded off, end-stopped, or fully defined. Nevertheless, in spite of their indeterminacy and because of

The Limits of Reason

their functional utility, irrational numbers are accorded conceptual status in the arithmetical system of thought.

George Lowell Tollefson

Unity

In arithmetical formulations, the number 4 is a symbol which stands for 1 plus 1 plus 1 plus 1. All numbers, formulae, equations, and operations, however large, small, imaginary, or complex, either directly or indirectly express such an operation. That is why the fundamental entity in this language is simple unity: the unit 1. Even irrational numbers are incommensurate approximations of rational numbers. For they are composed of arithmetical units, insofar as as they can be rendered into an articulate quantity.

The Limits of Reason

The Number Line

All arithmetical operations depend upon the convention of the number line. If 2 + 2 = 4 is proposed, its equivalent is: Ø Ø + Ø Ø = Ø Ø Ø Ø—i.e., two units added to two units is the same as four units. How is this known? It is known because the convention of the number line has previously been invented.

The number line (or counting numbers) is a convention which determines that, for any number in the sequence, the specific plurality of units in the number will always be followed by a plurality of units which is one unit more. For example, when the number 3, which is a plurality of three units, is augmented with an additional unit which is identical to any one of the three units, it will be the number 4.

And the set of units labeled 4 will then be followed by itself in conjunction with another identical unit, which new set of units will be labeled 5, etc. Thus, when anything outside the number line is counted, the count will always proceed in the same sequence as the number line: 1 2 3 4 5....

So, when identifying the number of a particular plurality of objects beyond 1, the accumulation of units representing the number of the plurality of objects is approached with this number line sequence in mind. Thus each whole number represents a specific set of identical units.

So in examining the equation, 2 + 2 = 4, a person first counts the units in the number to the left of the plus sign. She arrives at a number which is designated by the symbol 2. There are two units in this number. Then she counts units in the number to the right of the plus sign. She again arrives at a number of units which is 2.

The two counts are combined, as is indicated by the plus sign. So she counts straight through both numbers to reach four units, which count terminates in the number 4. In each of the three counts, the sum is labeled with the name of the last unit counted. Thus there is a 2 and a 2 and a 4. The four is the sum of the two 2s. So, since the two 2s are on one side of the equals sign and the symbol 4 is on the other side, it can be seen that they are the same because, when counting straight through the two sets of two units, 4 is the result. And that symbol is what is on the right side of the equals sign.

On neither side of the equals, or "same as," sign is the count terminated at 1 or 3. That is why those numbers are not specifically mentioned. For, though useful in the count, those portions of the number line are not under particular consideration and do not produce labels for the three elements in the operation: the two 2s on the left side of the equation and the 4 on the right.

Now the point of this discussion is simply to assert that arithmetical operations are not innate. They are the result of a practical invention, a convention called the number line. As mentioned in some detail in a previous work by the present author, *The Immaterial Structure of Human Experience*, the fundamental intuition in human experience is a sense of unity. It is this intuition which is formed by the intellect into a finite, and thus delimited, conceptual mold which can be applied to physical experience. As a result, discrete entities can be articulated by the mind.

An arithmetical unit (which is also the number 1) is an abstraction embodying the general character of a finite unity. The arithmetical unit is initially no more than an imagined finite uni-

The Limits of Reason

ty derived from the more vague but general sense of a unity. Prior to its use in an arithmetical system, it has no properties but those of finitude and unity.

This imagined unity is given a conceptual character, or logical utility, when, as an arithmetical unit, it is determined to be identical to all other such units within the arithmetical system. However, this is a conceptual, not a functional identity. For it may be altered in magnitude, as is the case between the rational numbers 1 and 1/10. As can be seen, there are ten 1s and one hundred 1/10s in the number 10. But within its particular context, be it as a whole number or as a particular type of fraction, the arithmetical unit remains the same abstract entity, resembling all other such units within that context.

Thus, in the context of whole numbers, it is of a particular magnitude. And, in the context of fractions, it is of another magnitude. In other words, all whole numbers are composed of varying sets of identical units. These are identical throughout the entire list of whole numbers. Thus 2 is composed of two identical units. And 3 is composed of three identical units, which are also individually identical to the units in the 2.

Likewise, all fractions are made up of identical units of a magnitude appropriate for the type of fraction. But the units in one kind of fraction, say the fourths in 3/4, may be identical to one another but different in magnitude from the units in another kind of fraction, say the fifths in 3/5. So, in many cases, the units of a particular kind of fraction will differ in magnitude from those of another and from the original component units of whole numbers. Nevertheless, they remain alike in that, within their particular context, they are units which are grouped with units identical to themselves.

Thus a whole number 3 is composed of three identical units. These are what have been referred to as the original units which compose whole numbers. So one third of 3 is also one of these units because 3 is composed of three whole number units. But one third of a whole number unit is not alike in magnitude to a whole number unit. And one fourth of 3 is not either. So these are of different magnitudes from one another and from whole number units.

Now take the example of pi. It is 3.14159.... The 3 to the left of the decimal is composed of three original whole number units. The digits to the right of the decimal are also each composed of units which are identical within each digit. But they are different in magnitude from the original whole number units and from one digit to another.

They progress by tenths in the direction of a lesser magnitude as they move further to the right of the decimal. And, as this progression is unending and continuously varying, it expresses the indeterminate number π. The indeterminacy is what makes pi a number which is irrational because it renders pi incommensurate with any rational number.

For a rational number is composed of a specific set of identical units, whatever the magnitude of those units in the particular number. Thus, even in the case of a fraction which is a repeating decimal (as in 1/3, which is .333...), there is a specific set of identical units. These units can be expressed as a ratio between two whole numbers, even if the decimal expression of that number is endlessly repeating. But, in the case of an irrational number, there can be no such specificity.

So, to sum up the point being made: Once the fundamental concept of an arithmetical unit is established, an expansion can occur in the stock of arithmetical concepts by initially creating a

The Limits of Reason

number line composed of identical arithmetical units. This number line is no more than an ordered set of sets of identical units, each progression from one number to the next number being a matter of adding one unit. That is to say, the identical units which compose the numbers of the number line can be grouped in a way which expresses a uniform order between the numbers.

This is the natural number line, which, when augmented by the concept of 0, is the basis of all the operations in arithmetic, including those involving imaginary and irrational numbers. For all numbers, insofar as they can be expressed at all, are expressed in terms of arithmetical units. This can be noted in the fact that even an irrational number, expressed in a non-repeating and non-terminating decimal, is set forth using integers from the whole number line.

George Lowell Tollefson

Proportion

At the root of the mathematical system is the phenomenon of proportion. An awareness of the possibility of proportion comes from the character of physical experience as apprehended by the mind. An encounter with experience involves a recognition of finitude. And this necessitates setting one thing against another in a recognizable pattern. It is in this search for a pattern that the mind seeks the rudiments of order.

So, for the mind to understand anything there must be order. Proportion is maximization of order. Thus even the most elemental organization of physical experience suggests proportion. So, where order is imposed upon physical experience, there must be a possibility of proportion as well, since any random order among physical extensions would suggest proportion to the increasingly order-maximizing mind.

However, proportional relations can be recognized in various ways. Consequently, the correlation between a mathematical system and physical experience, though essential for the mathematical system to work practically, is not initially set to specific relations. In addition, whatever relationship is established is limited by the fact that the quantitative proportions of mathematics are arranged in a predetermined pattern and cannot be altered.

The recognition of quantitative relations in physical experience is as follows. Individual mental impressions appear in human awareness in specific associations determined by their order of presentation to the mind. Many of these associations of

The Limits of Reason

mental impressions are identified as physical objects, the individual impressions constituting the qualities of those objects.

Other associations of mental impressions are thoughts, which have objects as their subject matter, however abstract the thought may be. The thoughts become increasingly systemized, modifying the mind's initial apprehension of the character of physical experience. Much of the more advanced portion of this intellectual activity is accomplished in terms of mathematical reasoning. And it is the consistent regularity of the mathematical system which serves to organize physical experience proportionately, insofar as this can be achieved.

George Lowell Tollefson

Prime Numbers

It is maintained in number theory that the square root of any prime number cannot be a natural number because a prime cannot have natural number factors other than 1 or itself. And neither 1 nor the prime number under consideration is the root. So neither is any other natural number the root. For it will not be a factor of the prime in question.

What is being focused upon is the manner in which the human mind creates concepts like those of number theory. For, like all of mathematics, number theory is a creation of the human mind. So, in attempting to understand the concepts of natural number, prime number, and composite number, one must have recourse to how the mind creates them. The first of these concepts belongs properly to arithmetic and has already been discussed in terms of the number line. Thus prime numbers and composite numbers are the concepts of concern.

Natural numbers are, of course, whole numbers not including 0. They are the counting numbers, which are a matter of convention. They are creations of the human mind from which complex arithmetical relations have been derived. These complex relations bring about a contradiction to the definition of natural numbers. For natural numbers are rational. And some operations with them produce numbers which are not.

But what is of concern is that natural numbers belong to arithmetic at large, while number theory is a narrow branch within that science. Number theory is especially concerned with the character of natural numbers (including natural numbers as negative integers). So, insofar as the properties of natural num-

The Limits of Reason

bers are to be considered, number theory offers rules which do not apply outside their unique conceptual domain.

The question which is of interest is this: are these number theory characteristics of natural numbers inherent in them? Or are they extraneous to them? In other words, are they an inevitable consequence of something deeply imbedded in the arithmetical character of natural numbers? Or are they not?

Perhaps they are the result of certain rules which are imposed on natural numbers only when they are considered under the auspices of number theory. If so, then the special characteristics of natural numbers, which appear when they are considered under number theory, would stand apart from the arithmetical definition of natural numbers.

As defined by number theory, all natural numbers other than the unit 1 are pluralities of arithmetical units which are either of a prime or a composite nature. For it is by means of this bringing together of multiple units that such numbers are considered plural. So the recognition of a plurality of units in any natural number other than 1 has been given a special emphasis.

A prime number is a natural number which has no factorable parts, other than the unit 1 or itself. Since the unit 1 or itself are not permitted as divisors, that means it is considered to be indivisible. Thus it is a plurality which is also a simple unity. For the plurality is held to be an indivisible unity.

Now a totality is the condition under which a composite number is considered. For a composite number is not a prime precisely because it is not a plurality of units which is held to be a "simple unity." Rather, it is a plurality of units which possesses factorable parts. But those factorable parts are not its arithmetical units. For, as in the case of a prime number, such a unit is not to be considered as a divisor of a composite number. As in the case of a prime number, neither the composite number itself nor one of its units can be considered as its divisors.

Yet it is because it can be factored that a composite number is not a simple unity. That is why, if a natural number is being referred to in number theory as a composite number, it is considered to be a totality. It is a totality of parts, not a simple unity, which has no parts. But it is not a totality of units. It is a plurality of units. Thus it is a totality of parts which are not its units.

So, when considered as a prime, a natural number is understood to be an expression of unity. It is not a totality because by definition it is not a unity of parts. Though it is a plurality of units, those units cannot be considered as parts. Parts are natural numbers other than 1 or the number under consideration. And a prime number must not be factorable into natural numbers. For these are not its divisors. That is what makes it a simple unity.

So a prime cannot be factored, as a composite number can, because such a factoring would be a contradiction of its stated character as a simple unity with no parts. However, that does not mean that the prime number 3 cannot be composed of three units. It simply means that, as a prime, it cannot be factored into parts other than those units. And those units are not to be considered as parts. That is, it cannot be factored into natural numbers other than 1 or itself. Thus it cannot have a natural number root either.

But this is why the prime and composite definitions of natural numbers are considered to be imposed conditions which do not arise out of arithmetical relations. For they are conditions which do not inherently belong to natural numbers, when they are considered simply as such. Without the rules of number theory, particularly the rule of the exclusion of units and the num-

The Limits of Reason

ber itself from the factorable definition of a natural number, primes would not exist. If primes did not exist, neither would composite numbers have any meaning.

Moreover, should a prime number be considered as a "totality" of units rather than as a simple unity of a "plurality" of units, the conditions imposed by natural number theory would dissolve. For, as a totality of units, it would be in character like a composite number and not like a prime. Thus, as a prime, the natural number 3 must be considered a simple unity. It must not be considered a totality, which would be to consider it as a composite of three units.

So, to reiterate, a prime number, like 3, is a number which is considered initially as a plurality. This is because it has multiple units. But then it is brought under consideration as a simple unity. The entire issue is one of deciding what the character of a natural number is. When considered under number theory, it is indeed of a character which is foreign to the normal operations of arithmetic.

For it can be seen that any natural number above the unit 1 is a plurality. So the question is, when should a natural number be considered as a simple unity or a totality? These are the considerations which define primes and composites. They are the conditions imposed by number theory.

George Lowell Tollefson

Composite Numbers

As evidenced in the previous essay, an investigation into the two most elemental concepts in number theory reflects a concern with how these concepts were originally conceived by the mind. It is not, however, an attempt to probe into strictly mathematical issues, which should be left to the professional mathematician.

What is desired is to determine what makes any branch of mathematics possible. Some of the elements of Euclidean geometry and the fundamentals of the number line have been discussed. So number theory cannot escape notice. For it is a foundational tenet of this book that knowledge can be fully accounted for in terms other than those of an innate conceptual origin.

In elemental number theory, the natural numbers appear to be composed of three conceptually distinguished sets. They are the number 1, prime numbers, and composite numbers. But how did such conceptual distinctions come into being? There can only be one answer. These three sets of natural numbers were classified in this way because each of them could either be understood to fall under a simple unity or a totality.

The unit 1, which is the number 1, obviously falls under simple unity. But take the concept of a prime number. It is immediately clear that a prime number is conceptually irreducible. That is, it can only be divided by 1 and itself. And, by rule, these are not allowed. So, having as divisors only 1 and itself, a prime number is understood to be conceptually irreducible because, as previously mentioned, it cannot be divided into parts.

The Limits of Reason

Since any consideration of number is here restricted to natural numbers, the parts would have to represent a natural number. This cannot be done with a prime number. So it has no composite parts. To state this more precisely, a prime number cannot be divided into a number of equal parts which are not the original units which formed the number. Thus it falls under simple unity.

Number in this context takes on a special meaning. For it has a restricted role in number theory. It designates a multiple of more than one unit. A number in this sense is any natural number which is greater than the unit 1, as in 2, 3, and 4. The significance of this designation lies in two facts.

First, a prime number cannot be divided into equal parts other than the unit 1. Thus a 3 is composed of three 1s, but has no other parts. And, second, the unit 1 is not considered to be a part. This is because a part is a number in the restricted sense that it must be a natural number greater than 1. It must be composed of more than one arithmetical unit.

So the unit 1 cannot be divided by a number to yield another number, since all numbers, as here defined, are larger than the unit 1. Nor can it be divided by another unit in such a way as to produce a quotient other than itself. Since it can only be divided to produce the unit 1, it cannot be divided into parts. Thus the unit 1 and all primes may be grouped together into a more inclusive set of natural numbers which falls under simple unity.

So, to account for composite numbers, the concept of plurality must be employed in a manner different from that of the previous essay. For a composite number is composed of a plurality of sets of multiple units. Take the composite number 150. It is composed of either seventy-five sets of two units, fifty sets of three units, or thirty sets of five units. Accordingly, a "set of multiple units" indicates the original plurality which distinguished the number of units in a prime number in the previous essay.

But, in the present case, the concept of a plurality has be altered to refer to multiples of sets of units, rather than to multiples of individual units. It is these sets which are defined as numbers that are greater than 1. Unlike the case of prime numbers, the sets of multiple units are, when considered collectively, recognizable as the elements constituting the plurality of a composite number. This plurality is always a multiple of sets, each set representing the same number, though there may be different pluralities in the same composite number, as indicated above.

There is no plurality of sets of multiple units in a prime number. There is only one set: the prime number itself. Thus it is by means of a plurality of equal sets—which can be recognized as multiples of a particular number and therefore labeled as equal parts—that a composite number can be defined as a totality of equal parts.

It is for this reason that a composite number is not a single irreducible set of units. If it were, it would be a prime. And the plurality of the units could be ignored, the set itself being treated as a simple unit: a prime number. This is because the issue of a plurality of units has been put into abeyance by a shift in the meaning of "plurality" to *multiple sets* of units.

The restriction to multiple units in the definition of number results in the fact that the unit 1 cannot be considered a number. Therefore, it is not considered a part. For parts are numbers. Moreover, when plurality is understood as being significant in the recognition of composite numbers, it is no longer of principal importance in the recognition of prime numbers. For they

The Limits of Reason

are looked upon as simple unities. Thus the term "plurality" has taken on a new meaning. And it is that new meaning which supports the governing concept of a composite number: totality.

Thus the composite number 6 can be a totality comprised of two numbers. Or it can be a totality comprised of three numbers. In other words, the composite number 6 is made up of equal parts which may be identified as either the number 3 or the number 2. Since it may be one or the other, it has two prime factors.

Again, the composite number 12 can be comprised of two 6s, three 4s, four 3s, or six 2s. But the first two numbers, 6 and 4, may be further divided by natural numbers greater than 1. And these divisors are the second two numbers, 3 and 2. The numbers 3 and 2 cannot be further divided by a natural number greater than 1. So they are the prime factors of 12, just as they are the prime factors of 6. Consequently, the multiple equal sets of 12 can be either four 3s or six 2s.

It is in this way that prime numbers can function as the equal parts of composite numbers. They act as simple units, somewhat resembling the unit 1, inasmuch as they are conceived as units. It is these which combine as equal parts of the totality which is a composite number. Thus the composite number falls under the designation of a totality.

Unlike the prime numbers and the unit 1, any composite number can be divided into parts. That is, it can be factored into natural numbers which are indivisible by other natural numbers greater than 1, and which are therefore prime. As can be seen from the examples of the composite numbers 6, 12, and 150, these parts are equal sets of multiple units.

But to further demonstrate how what has been discussed represents a separation into equal parts, multiplication must be reduced to its underlying structure, which is addition. Multipliers are simply a convenient shorthand for referring to the addition of identical addends. They are the number of those addends. Thus it is evident that all composite numbers are sums of identical primes.

So, among the fundamentals of number theory, the principal questions to be asked are, first, whether or not a natural number can be evenly divided by another natural number greater than 1. Second, if so, will this quotient express a natural number greater than 1? And, third, will that number be indivisible if approached in the same manner? The critical point to be made is that what are being referred to are "numbers," not the unit 1.

In the language of this system, the unit 1 is not included as a number. This is what is meant when it is said that primes are irreducible, that they cannot be divided into parts. What is meant is that primes cannot be divided into equal natural numbers greater than 1. But composite numbers can be reduced into equal natural numbers greater than 1.

This conceptual specialization in the unique case of number theory is what renders the fundamental concepts of the unit 1, prime number, and composite number. So, since numbers greater than 1 are held in this system to be either primes or composites of primes, it can be seen that unity and totality are the operative concepts of the system of natural numbers insofar as the fundamentals of number theory are concerned.

The Limits of Reason

The Infinitude of Primes

It is not the mutually exclusive character of the terms "finite" and "infinite," but rather the mutually exclusive character of the terms "determinate" and "indeterminate" which accounts for Euclid's proposition IX.20[31] concerning the so-called infinitude of primes. For this reason, the use of the concept of infinity in this context should be questioned.

Because it is terminally bound in both directions according to its definition, a rational number is a finite entity. Consequently, the number 15 is distinct from the numbers 14 and 16. It is a finite entity, like a grain of sand. And every individual fraction, however large or minute, is a similarly bound number.

So, since finite parts can only make a finite whole, as grains of sand make a beach of perhaps great but not infinite extent, it follows that any set of rational numbers, however large, is finite. For, just as any number of finite grains of sand can only compose a finite beach, it is equally true that any quantity of finite parts can only make a finite whole.

In this way, a beach composed of an unimaginably large quantity of grains of sand does nevertheless encompass a finite quantity of sand. The unimaginably large quantity of grains of sand may be indeterminate. But it is nevertheless finite. It is not infinite. That is to say, it is not *not*-finite.

It follows then that the set of rational numbers is finite. For it is composed of finite entities, however innumerable they may

---

[31] *Euclid's Elements*, Book IX, Proposition 20.

be in the practice of accounting for them. And any set of primes, being a subset, and therefore a part, of the set of rational numbers, is also finite. But Euclid, in proposition IX.20, is thought to have proven that the number of primes is infinite. Yet he does not assert this. He merely demonstrates their indeterminate character.

So, because he is thought to have proven that the number of primes is infinite, the above statements appear to contradict him. However, Euclid must be considered in terms of his logic and not by any interpretation of his meaning which may be attributed to him by others. What he demonstrates is that for any set of primes, however great in number, there will always be another prime number.

How many rational numbers there are is simply not known. But it can be seen that individual rational numbers are finite, or determinate. For the definition of a rational number, as opposed to an irrational number, is that it be finite, or determinate. If these numbers are finite in themselves but indeterminate in quantity, it may be assumed that the consideration of all rational numbers leads to a finite but indeterminate result. Thus the set of all rational numbers is indeterminate. But it does not follow from this that the set is also infinite.

Therefore, for the above reasons, it can be concluded that any set which is composed of finite, or determinate, entities is itself a finite set. This means that a set of finite entities, such as a set of rational numbers which is presumed to extend to some final count (at whatever point that count may be taken), must by its very nature be determinate, even if it is not known what the final count is. The fact that another entity can always be added does not matter any more than the fact that another stone can always be added to increase the size of a pile of stones. One

The Limits of Reason

would not assume from this that the pile of stones can be made infinite.

This is because determinateness is precisely what is being considered. Thus the same applies to the set of prime numbers, which is a subset of the set of rational numbers. For, since the latter must be presumed to be determinate, or finite, in character, even if it is indeterminate in practical terms, the former is also determinate, or finite, in character, though its practical determination is indeterminate.

However, what is being spoken of is not things but concepts. The indefinite concept of all rational numbers is therefore indeterminate as a concept. For it is imprecise in imaginative reference. That is, though it is defined, the imagination cannot supply all of its particulars in terms of a concrete image. For all its members cannot be set forth. So there is not a complete knowledge of what is included within the set, though both the finite character of each of its members and the appropriate steps for increasing the number of members are known.

Consequently, by logical extension, since it is not known how many primes there are, the indefinite concept of all primes is indeterminate. Moreover, a discussion of such indeterminateness cannot be reasonably concluded under a heading involving determinate entities. For the concept is *in*-determinate and lacks the precision of a determinate concept.

In other words, a conceptual indeterminateness cannot be logically derived from a discussion involving determinate concepts. Such a line of reasoning can only be carried out as a means of pointing from what is to what is not. Thus an indeterminateness is not a determinateness. As existence suggests non-existence by means of the imagined removal of the existing thing, so determinateness suggests indeterminateness by means of the removal of the definitive bounds of the determinate. So, since it is only a suggestion, it leads to the obvious conclusion that the indeterminate is, as a conception, excluded from the domain of the determinate. No more can be said of it.

The problem lies with the concept of a set. A set is a logical grouping, or classification. If all of its content is not known, then the range of the set is not known. It has an indefinite range and is therefore an indeterminate set. To declare otherwise is to state that something which is unknown is in fact known. That is a contradiction. Or, in Euclid's terms, if there is always one more prime in the set, the set cannot be closed in a determinate manner. It is thus an indeterminate set. But this does not mean it is not finite.

The Limits of Reason

Mental Focus

Addition and subtraction are the fundamental operations underlying numerical science. Addition has been considered. So let the operation of subtraction be examined. Visualize five apples laying on a table. Remove two of them. Three are left. This phenomenon is not only recognizable in imagination, but is immediately familiar to experience. So it can be seen that it was from empirical phenomena that the principle of subtraction was initially taken.

Its conceptual realization is expressed in an equation like $5 - 2 = 3$. Such a statement is an abstraction from physical experience. Of the original empirical content, all that appears to remain in this equation is a relation of units which can be applied to any class of objects that is arranged in accordance with a numerical progression. Thus two units (the number 2) are removed from five units (the number 5). And three units are left (the number 3). This relationship was found in something like apples. In other words, it appears to have been discovered in physical experience. But not the concept of number.

Now among the most fundamental modes of mental awareness are those of unity and division. Consciousness itself is a unity. In fact, it is the foundation of any sense of unity. For all things apprehended within it are held in a unity within it. This is, in effect, the means by which the intuition of simple unity is exercised.

It is also the manner in which the intuition of plurality is exercised. For there cannot be plurality without the recognition of unity. Plurality is separate unity. And of course totality follows upon these. For the intuition of totality arises from the recognition of a plurality of separate unities.

So, if five apples are focused upon by the conscious mind, they are held in consciousness as an independent totality. But whenever an attempt is made to distinguish two apples from the five, a division occurs. This division is a breakdown of the original totality of five into the two lesser totalities of two and three. This division is not arithmetical division, but subtraction.

Thus the latter two totalities, two and three, are portions of the original whole. And there may be portions of these portions. For the two apples and the three apples may each be divided in this way again. The two apples may become one apple and another apple. And the three apples may become two apples and one apple. Or they may become three separate apples.

Now, returning to the table supporting five apples, not only are the table and apples perceived. But other nearby objects within a limited range of mental focus can also be included. Recognition of greater or lesser groups of individual objects (perhaps apples, a table, and a wall) results from various levels of mental focus being applied to experience. Moreover, because this is a free mental process, what are recognized as objects can also be modified by the mind. Thus an apple can be broken down into a stem, seeds, core, flesh, peel, etc. Or it may simply be recognized as a globe of fruit.

In any case, these objects become specific expressions of unity within the larger totality of an overall perception. For, though objects must be apprehended in terms of unity, the focus of that unity can be shifted. Consciousness may enclose within its sphere of attention, and thus within awareness, a smaller or greater field.

The Limits of Reason

So its focus may rest on only a part of an apple, a whole apple, two apples, or five apples excluding the contiguous table and other proximate objects, like the wall behind the table. Or it may include those objects as well in one general unity of focused awareness: a totality of objects within the field of vision.

So all these objects inclusively are one unity: a totality. And within a more restricted field of focus are five small unities, which together are the lesser totality of five apples. In other words, these five apples are a totality within the greater totality. Within this lesser totality are two, three, or one apple. Thus the mind can focus upon five apples as a unity, then upon two apples as a unity within that unity, set them aside, and observe what is left, a unity (or totality) of three apples.

The final count of three apples results from a focus upon each of the three apples individually. The same had been done with the two apples. This may also have been done with the original five apples, though it was not necessary for the operation to take place. It is only necessary to identify the five apples as a general multiple of apples. The counting occurs in the operation of subtraction. It is in this way that the flexibility of mental focus can be seen to permit such an operation, which is immediately accompanied by tallies of the subtrahend and difference.

As already demonstrated, two individual apples, focused upon as a group, are tallied as two. They are removed from a larger group of apples. What remains is tallied as three. It is then noted that the original group is tallied at five, when all are counted together. So this final tally accounts for just how many individual apples were involved in the process.

It is thus acknowledged that the remainder of three apples had belonged to a specific larger class: that of five. In turn, this largest class of apples, which was the content of an overall focus on the original totality of apples prior to any shifting of that focus to fewer apples, had itself been a result of a division between two classes of objects—the table and wall in one and the apples in the other. The latter, once isolated by means of mental focus, is then further divided into two and three objects respectively, according to the operation of subtraction and its result.

So the controlling factor of any mental operation is a field of conscious awareness, be it an awareness of concepts, images, or things. For all these take place in the mind. That field of awareness may be either narrowed or enlarged in focus. A narrowing of focus is what happened when attention was moved from the apples, table, and wall to a concentration on the apples alone. It is also what happened when attention was moved from the five apples to groups of two and three.

Consciousness, then, is unity. And attention to a smaller domain within a larger field of awareness is a prerequisite to the operation of subtraction (i.e., analysis into parts). When attention is narrowed within a domain of five apples, two smaller domains of two and three apples are consecutively recognized.

So far, there has been an alternate shift in play between comprehensive and subordinate unities, which latter are themselves totalities of individual unities. In this way, a field of awareness and a narrower domain of focus have been established in contradistinction to one another. Accordingly, it is by an act of will that the mind can narrow the scope of mental awareness to take up a position of reduced focus.[32]

---

[32] Again, it should be pointed out that this discussion is from a representational viewpoint. From an immaterial viewpoint, the action of will can be seen to lie much deeper in human experience. For it lies in spirit, or univer-

The Limits of Reason

From this point, it can move its focus into an act of further narrowing within the adopted field of awareness. The human will thus makes use of the manner in which the mind works through modes of attention, or focus. So it is in this way that the isolation and identification of objects in experience is accomplished. When accompanied by number, it is a quantitative identification.

Concepts may be representative of things like trains, cars, and persons. Or they may be abstract, as in the equation mentioned at the beginning of this essay. The general character of mathematical concepts is an expression of the mathematical system as a whole. But such abstract concepts may be rendered practically useful by being applied to thoughts of physical things like those mentioned at the beginning of this paragraph. Or they may be applied to other abstract thoughts.

The mathematical system was created in the mind. But, in the matter of its potential regarding the relations it expresses, it originated among observed patterns in physical experience. Nevertheless, the further development of complex relations in a mathematical system is independently elaborated in the mind. For it follows a logic not generally recognized in the simple proportions of physical experience.

So it is in this way that a means of applying mathematical patterns to physical experience is instituted. The patterns are recognized as logically necessary within the mathematical sys-

---

sal consciousness, beyond the material expression of space and time. Thus all the acts of that will, as regards an individual person, are concurrent and instantaneous. And the presentations of phenomena in the mind, including variations of mental focus, would appear as if predetermined. For an understanding of the immaterial viewpoint, see *The Immaterial Structure of Human Experience* by the present author.

tem. For they are developed through the conceptual associations of that system. Subsequently, when successfully applied to frequently encountered consistencies in physical relations, they become laws of those relations.

But behind this entire process remains the fundamental mechanism which is a shifting of mental focus in relation to a field of awareness. Such a shifting of mental focus within or beyond a particular field of awareness results in either smaller, more narrowly focused, or greater, more broadly focused fields of awareness.

A physical law backed by mathematical relations is felt to be universal, since it originates in the logical structure of mathematical thought. Its universality is asserted when it can be applied anywhere within a field of awareness in which the relevant physical relations are encountered. Yet the fact remains that the potential for establishing the mathematical relations in the first place was discovered in physical experience.

This discovery was the observation of relations of greater and less in regard to something like apples, as was discussed at the beginning of this essay. Thus mathematical rigor does not exhibit innate relations of the mind. The mathematical system is invented. But, at the same time, its development is dependent upon a faculty which the mind introduces as an organizing agent for experience. That organizing agent is the intuition of simple unity. This intuition is an exercise of consciousness and focus.

The conscious mind has used its control over its faculty of focus as a means for producing abstractions which have been worked into a complex system of mathematical thought. It has done this by extracting the uniform property of the arithmetical unit from the wide differentiation in relations it has encountered

The Limits of Reason

between physical objects in experience. But, in its creation of a number system from that arithmetical unit, it has greatly elaborated on this property independently of physical experience.

Thus the principle of number (though not the fact of it) may be held to inhere within physical experience, in which order and proportion are sought and subsequently found. Number, understood in this way as corroborating order, may be considered to be systemized proportion. So the five apples, two apples, and three apples previously described in this essay were initially not quantified perceptions. Rather, they were orderly extensions of greater and less. And, as such, they lay within the domain of physical experience. But the equation $5 - 2 = 3$ lies within the domain of reason.

George Lowell Tollefson

The Binary Mind

Even numbers are binary for the simple reason that the number 2 is inherently so. For the evenness of any number represents a pair of equivalent entities. But the question to be broached here is, does this suggest something deeper? Is there a binary character which inhabits all number operations because the human mind must process numbers in pairs and in no other way?

To simplify the discussion, let it be acknowledged that multiplication is an abbreviated form of addition. And addition is always a matter of combining two numbers at a time. For example, in the equation $2 + 1 = 3$, there is a single binary operation combining the 2 and the 1. But this equation is expressed in an abbreviated form. For the first act of addition has already been performed. Two 1s have been combined to make a 2.

In the extended version of this equation, which is $(1 + 1) + 1 = 3$, it can be seen that the two plus signs indicate that the numbers submit to a double binary process. First, the two 1s within the parentheses are added together. Then another 1 is added to the previous sum. What this demonstrates is that the universal underlying foundation of any additive process is that it proceeds in a binary manner. In fact, this is true for any mathematical or mental process. For the mind is a binary instrument.

Now a reason for the binary role in multiplication is this: Multiplication, like addition, is an operation which seeks to combine units. But the human mind can only clearly conceive a relationship between two things at once. So, if it is asserted that $2 \cdot 4 = 8$, what is being stated is that two 4s are 8, or $4 + 4 = 8$.

The Limits of Reason

This is a single binary operation. But if it is asserted that $4 \cdot 2 = 8$, what is being stated is that four 2s are 8, or $2 + 2 + 2 + 2 = 8$. These numbers are added together incrementally in pairs, as in $\{[(2 + 2) + 2] + 2\} = 8$.

These incremental operations are binary in character. In the bracketed formula $\{[(2 + 2) + 2] + 2\}$, 2s are being incrementally combined, beginning with the number 2, progressing by a 2 to 4, then by a 2 to 6, then to 8. In each case, one number is combined with another. Thus the numbers progress to larger quantities two at a time. So there are two elements being dealt with at each step. The multiplication is simply a memorized shortcut for performing this tedious operation in addition. It can be seen from this that multiplication in general masks an incremental addition of pairs of numbers in successive stages.

In the case of division, the reduction will follow the above pattern in reverse in the manner of subtraction. The equation $8 \div 4 = 2$ asks, how many 4s are in 8? So a 4 subtracted from 8 leaves a 4. Since there are only two 4s, this represents a single binary operation. But, in the case of $8 \div 2 = 4$, the equation asks, how many 2s are in 8?

This does not represent a single binary operation. Rather, it suggests several successive binary operations. In other words, the number 2 must be subtracted incrementally in pairs of numbers which proceed from increasingly smaller quantities. This is the operation $\{[(8 - 2) - 2] - 2\} = 2$, which subtracts 2 from 8, then 2 from 6, then 2 from 4 until a final difference of 2 is arrived at.

Thus again a pair of numbers is the focus of an independent operation at each step. So it can be seen that division is an extension of the concept of subtraction in the manner that multiplication is an extension of the concept of addition. It can also be seen that this subtraction undergoes binary procedures.

For example, divide something in half. The result is a duality of two parts, as in $1 \div 2 = 1/2$, in which there are two halves. But divide something into three parts, as in $1 \div 3 = 1/3$. And the result does not seem to indicate a duality. For the mind appears to be processing three parts at once. And 3 is more than 2. But, in spite of appearances, this operation must be carried out in a binary manner. One part is separated from the whole. Then another part is separated from the difference, leaving a final part.

In other words, to arrive at the three parts, two consecutive cuts must be made, as in dividing a pie. These consecutive cuts resemble subtraction, each involving two numbers: a minuend and a subtrahend. This reveals the fact that division is subtraction. It also reveals the fact that, as in addition, the operation of subtraction must be performed consecutively on pairs of numbers.

So, to explain these mathematical processes, only two operations are necessary, both involving forms of mental focus. They are unity and duality. Duality is the basic form of plurality. So the employment of these operations reflects the intuitions of simple unity and plurality. Simple unity is an intuition which is derived from the experience of pure consciousness and its capacity to encompass and unify a field of awareness.

The intuition of plurality is initiated by the experience of a diversified content of consciousness. This diversity is apprehended two elements at a time. When a person sees or imagines three entities, he envisions two, then that pair and the other entity. This generally occurs so quickly it is thought that the three entities are apprehended at once. So plurality beyond duality is simply an iteration of the practice of identifying dualities.

Probability

For the human mind, there is no such thing as chance. Chance is a word which has no comprehensible meaning beyond negation. It is a word that universally denies order. Were there such an absence of order, there would be chaos. And chaos is an equally empty concept, itself devoid of any meaning other than that of negation. It simply means the complete absence of order.

A complete absence of order is something which the human mind can neither grasp nor accept. For the human mind shapes its map of experience by giving it order. Thus it can be seen that order is a condition of human awareness. A conceptual grasp of that order, however perfected or flawed, is the only work possible for the human mind.

So, if chance cannot be spoken about meaningfully, probability can be spoken of intelligibly. Probability might be called the science of chance. But this is not universal chance. It is not chaos. It is the throwing of a die or the arrival of different kinds of weather. A specific range of outcomes is expected. But precisely what that outcome will be at any particular instant is not known.

Thus, assuming the mind's need to both recognize and establish order in what it observes, an intelligible explanation for any event is required, even if that order is incomplete. For this reason, changing phenomena must have predictable results. Precisely what a result will be may not be immediately recognizable. But some reasonable range of expectation must be assumed.

For a recognition of causal relations is required if any rational sense is to be made of events.

Let it be assumed that an event is about to take place. But precisely what event is not determined. It may be included in a set of several possible outcomes. Yet it is nevertheless expected that this particular set of outcomes will be the limit of possible outcomes. There are a good many more outcomes which are not to be expected. These kinds of considerations are what place events into an orderly progression. They are necessary if a prediction is to be made, be it directly causal or more loosely probable.

To begin, an antecedent to any predicted outcome must be found. For the expected train of events must start somewhere. But, once an antecedent is established, it will lead to a direct or probable result. Otherwise no rational sense could be made of the progress of events. So some sort of order must be recognized.

Intelligence demands explanation. And, in matters of behavior or decision making, anticipation must find a path to go forward. For the anticipation to be resolved in action, there must be a direct cause-and-effect relationship or its nearest equivalent, a relationship of cause to a probable array of effects. For these are forms of relations which inhabit the orderly channels of the mind's necessary functioning.

Thus, when one event follows another event in a predictable relationship between two events, the two events are causally linked to one another. If they were not, they would be causally dissociated from one another. But, if dissociated, then they would nevertheless be associated with other causally antecedent or consequent events. For, if they are to be thought about with a

The Limits of Reason

possibility of prediction, they must follow some sort of causal order.

This is because past, present, and future are generally assumed by the mind to be a seamless whole, an underlying structure for a universal order expressed in the sequences of change, or time. Even if an inexplicable event, say a miracle, should be experienced, it would soon be provided with an explanation, an arrangement of antecedents and consequents, and made to fit seamlessly into an enlarged fabric of the whole.

Probability follows this causal progression, albeit a little more loosely. For probability does assume order. But it also acknowledges that it does not have fully within its grasp a sufficient supply of the details of that order. For this reason, insofar as its probable relations are concerned, its working environment must be treated as open-ended.

It cannot be treated as a closed box, within which all the contents are registered in the mind. For the precise order of some of those contents remains enshrouded in mystery. However, it is nevertheless assumed that, if all the details were to be known, its working environment could be understood as a closed box. Otherwise, probability would have no determinate outcome whatsoever. Nothing could be expected of it. Nothing would be probable.

This is true even in quantum science, where anomalous relations in particle behavior make their appearance. The anomalies cannot be rectified under present observation. So definitive causal explanations must be set aside in favor of statistical explanations. But, in spite of arguments to the contrary, the fact remains that the probable relations imply an order that is not fully revealed. For the mind is ever in search of order. That is why the statistics are being used, even if it is assumed that certain causal irregularities can never be resolved.

It can be said that probability presumes a sort of suspended determinism. There is a definite web of causal relations to be reckoned with. But some of the fibers of that web appear to be missing. However, it is understood that they are missing in appearance. At a deeper level, perhaps at some future time, a better informed reckoning will be made.

If this cannot be done, a probability is better than no sense of order. So the crux of this discussion has been that the science of probability carries within it an element of chance only because something which would be expected to determine an outcome is left unspecified. The outcome is therefore partially undetermined.

Now let an emphasis be placed on a variation in cause rather than its outcome. Several causes may individually produce the same outcome. When the specific cause is not known, this also becomes a partial causal relationship. Though the outcome may be ascertained, there is a probability as to what brings it about.

In any case, a probability is not a causal relationship between known antecedent conditions and their known outcome. For, if any portion of the cause is unknown, neither can its connection to a specific effect be affirmed. Likewise, if any portion of the effect is unknown, neither can its connection to a specific cause be affirmed. There is always an element of doubt as to the relation. It is this sort of limited causal relationship which determines what is commonly meant by the word "chance." It is a probability.

For this reason, all that is known to the human mind is that things must appear to be more or less determined in their relations. This is enough for the mind to reason about events and

The Limits of Reason

make predictions. The predictions may have a probable or certain outcome, depending on how much is known of the antecedent and consequent conditions. Beyond this, there is only faith in the uniformity of experience. Such faith is characteristic of the human mind.

George Lowell Tollefson

Commensurable Relations

Just as the tenth book of Euclid's *Elements* introduces a theory of incommensurables which are resolved in the square (what would now be called algebraic irrationals), so are more subtle incommensurable relations tucked into the definitions throughout the *Elements*. The incommensurate character of the definitions is kept out of the propositions, except insofar as the definitions as a whole are directly or indirectly introduced into their proofs.

What this means is that no attention is drawn to the fact that some of the definitions are incommensurate within themselves. Others are incommensurate with one another or with the physical world. An example of the former, as introduced in previous essays of this work, would be the incommensurate relationship between the diameter and circumference of a perfect circle. An example of the latter is the definition of a line as a "breadthless length."[33]

The first example is a relationship which can only be expressed by an irrational number: a transcendental number which is incommensurate with any other number. The second example, the definition of a line as a "breadthless length," is commensurate with absolutely nothing in the physical world.

It is likewise not commensurate in figurative terms with other Euclidean definitions, such as that of a point. For it can be

---

[33] *Euclid's Elements*, Book I, Definition 2.

The Limits of Reason

asked, how can that which is defined as having "no part"[34] be figuratively imagined to be located upon that which has no breadth? This cannot be represented in the mind in a concrete, or visualizable, manner.

It is true that Euclid's line is a concept drawn from experience, as in the recognition of an apparent edge on a box, a sawed plank, a molded brick, etc. As such it is a boundary, which is a meeting of two surfaces, say those of the brick and its immediate surrounding environment. But Euclid's line is rendered independent of this type of physical experience through the mental process of abstraction. For no indisputable, concrete models of the concept of a breadthless line can be found in nature. This is because a meeting of two surfaces is not an identifiable breadthless line. The line is simply an arbitrary reference for marking a differentiation between two surfaces.

So there is an incompatibility of the concept with anything other than itself. But let it be candidly pointed out that, where an incompatibility is being referred to as an incommensurability, the concept of incommensurateness is being extended into unfamiliar territory. Some might say the meaning of the concept has been attenuated. For mathematical commensurability is a concept that involves a ratio between integers (other than 0) which yields a rational number. Thus an incommensurateness of ratio would yield an irrational number.

But what has just been introduced is an incompatibility between an abstraction and an object in the physical world. No quantitative reference to either the abstraction or the object has been made. Yet the incompatibility between them has been lik-

---

[34] *Euclid's Elements*, Book I, Definition 1.

ened to an incommensurate relationship. So are incompatibility and incommensurability really the same thing?

Yes they are. First, let this be examined in terms of the physical. Physical incompatibility involves a disproportion. It is a disproportion in properties. In other words, if one speaks in terms of proportion, one rock might be considered compatible with another in its size or its mineral content. In either case, a measurement may be made: one size to another, or one quantity of mineral content to another. A proportion is established. That is to say, if a measurement of any use is to be made, it must establish a proportion. Otherwise, a measurement cannot be taken.

But the concept of an ideal circle is different. It is incompatible with a circle in the physical world because no such proportional comparison can be made between them. Such a measurement cannot even be considered. The same is true of the breadthless line. How is one to make a comparison of properties between the concept of a breadthless line and a line (or boundary of an object) found in the physical world?

Both comparisons present the same problem of incompatibility. In each case, it can be asserted that the incompatibility is in fact an incommensurability because to state that there is no possible measure of comparison is to imply that there is no workable proportion. That is, no rational quantitative outcome from such comparisons can be expected.

A clue to this problem now seems to have become obvious. It lies in the apparent fact that only the physical can be compared to the physical in terms of measurement. But is this true? It can be seen that measurement is a matter of determining some sort of proportion, which is a mathematical relation. And a mathematical relation is a conceptual, not a physical, matter.

The Limits of Reason

So it is beginning to appear that the act of measurement is both mental and physical. This means that, for the measurement to take place, there must be something in the mathematics which corresponds to the physical. Something in one that is in the other as well. But it is not mathematics as a whole which expresses this correspondence.

It has already been observed that mathematics is conceptual. Not only is it conceptual. It is conceptual in a rather elaborate way. It is an abstract system related only to itself. That would place it in a realm entirely separate from the physical, were it not for the fact that its internal relations are proportional. Physical relations are also susceptible of proportional arrangement. They can be made to conform to mathematical proportions.

So it seems that proportion in some sense lies at the foundation of both thought and the physical. Not just mathematical thought. All thought. "All" thought is included because, when the rock's weight or mineral properties were being compared above, the rock had been translated into a mental concept. It is in this manner that any proportional comparison between physical entities is made. The entities are translated into concepts. This is what allows them to be brought under the aegis of mathematical concepts, such as those of number and proportion.

Since it is true that the internal relations of mathematics are relations of proportion, it can be seen that two mathematical concepts, as in two concepts of measurement (say the numbers 10 and 12), can be proportionally compared to one another. But, if this comparison is made, it cannot be maintained that they are being physically measured. They are simply being associated by means of their definitions.

The common element between the two numerical concepts 10 and 12 is the arithmetical unit. But what exactly is an arithmetical unit? Is it the number 1 or the 1 in the number 1/10? And what, for that matter, are these two numbers 1 and 1/10 if they do not involve the same magnitude of unit? What are arithmetical units? And what are numbers?

These are not physical questions. For it is the nonphysical, logical character of number which provides an association between the two concepts 10 and 12. The association is being made in terms of units. A unit is not a thing, like a rock. It is a concept which is variable in magnitude. So it is either a 1 in 10, of which there are ten units. Or it is one of twelve units in the number 12. But a question arises when one considers what the number 1/10 is one tenth of. If it is one tenth of 5, the units are half the size of each of the five units in 5.

In the case of the numbers 10 and 12, the arithmetical unit is designated to be the number 1. There are ten such units in 10 and twelve such units in 12. But in one tenth of 5, the unit would be one half of the number 1. Whereas, if the unit is one tenth of 1, then the number 5 has fifty of them in it. Consequently, the unit itself is clearly immaterial. This is demonstrated by its arithmetical property of taking on differences in magnitude, as in the unit being either a 1 or a half or a tenth of that 1, depending upon the logic of its employment.

Its only property is that it is consistently the same in magnitude for a particular numerical comparison, such as the comparison of the numbers 10 and 12. That is how it can be known what is meant when 10 and 12 are compared in the mind. Other than to speak of both numbers in terms of the same magnitude of the units of which they are composed, it could not possibly be said what these numbers are. For they exist comparatively. And that comparison is made between their constituent units. The number 10 is composed of ten identical units. And the

The Limits of Reason

number 12 is composed of ten units identical to those in the number 10, along with an additional increment of two such units.

The concept of a number is as nonphysical, and therefore as difficult to visualize concretely in the mind in relation to all other numbers, as are the arithmetical units of which each number is composed. The difficulty arises from the differences in magnitude to which the arithmetical unit is susceptible. Hence a .10 or .12 is different in magnitude than the 10 or 12 just mentioned. However, there is always a uniformity of the unit within specific magnitudes, however complex the relationship between the numbers composed of them may be.

On the other hand, physical things are concrete. A ten inch brick and a twelve inch brick can be compared in terms of qualities which are independent of number. Generally speaking, one is thicker, longer, or heavier than the other. These properties are recognized as involving direct mental impressions prior to their incorporation in concepts.

But they can only be quantitatively compared by measuring each with a separate physical standard of measure, say a ruler or a weight scale. Yet how was the ruler measured or the scale graded? By another physical object or mechanism. Should this regression be pursued, it would continue indeterminately. So does that mean the instruments of measure are as concretely indeterminate as the numerical concepts themselves?

Of course not. The leap from a concept of number to the physical use of it is inherently incommensurate. Numbers are not things. Though a counting sequence may be applied to things in such a way that they may be *said* to have number. Nevertheless, numbers and the things counted with them are intrinsically incompatible. For numbers belong to an independently integrated system of concepts. This system of concepts has no integrated relationship with those things upon which it is used to form a count.

In other words, the counting process follows rules which define the relative meaning of the counting numbers without imputing the character of number to the physical things numbered. A fifth stone is a fifth, not in stones, but in the counting series of numbers. Otherwise, it is just a stone like any other. In fact, it is a stone unlike any other when it is examined closely.

Likewise, the fifth numerical gradient on a ruler is a counted mark, subsequently so labeled. It depends on the number of prior marks for its meaning—i.e., those prior marks as submitting to a counting process. Otherwise, it is just a mark like any other (or, to be minutely precise, unlike any other). Thus there is an incommensurateness between any numerical concept and the object to which that concept is applied. But for practical purposes the incommensurability is conveniently ignored.

So far, the observation that all comparisons imply measure has been adhered to. Measure indicates proportion.[35] The only apparent exception to this rule is that in practice, in many cases, it is often not recognized that comparisons of objects have to do with proportions. Qualities and properties are more vaguely spoken of instead.

For example, a person may state that this quality is blue; that quality is red. Or he may observe that this metal has the property of possessing great tensile strength; that does not. One need not measure the quality or property by degree. A general com-

---

[35] A note to the reader: the word proportion is frequently employed to mean ratio.

The Limits of Reason

parison will often do. It is a matter of how precise one wishes to be in expressing a comparison.

That a dog is defined as being warm-blooded does not at first glance appear to have anything to do with measurement. A further analysis is required to supply the precision. To achieve such precision, the character of warm-bloodedness in animals must be further explicated. For example, a person may consider the degree of internal temperature regulation. Thus, where precision of measure is omitted, it is implied. Even comparisons of such qualities as shades of blue are comparisons of degree, either implicitly or explicitly.

Accordingly, since thoughts about physical experience are centered upon mental objects, which is to say that the physical objects are held within the thoughts as objects for mental consideration, it should not be surprising that an emphasis on proportion is true both of comparisons between one image and another and between one concept and another. For the physical objects of thoughts, both considered apart from the thought and considered within it, are composed of properties. And the properties are associations of individual mental impressions.

An image is imaginative. It is somewhat careless of the details, even in its implication of them. So it only suggests a comparison. The comparison of concepts, on the other hand, is explicit and deliberately precise as to details. For it is assumed that an exact comparison of the details can be made. Consequently, a concept comparison can contribute to logical implication. This is often a broadly proportional comparison of properties: all, some, none, or not all. But it is nonetheless exact in implication.

This common characteristic of a comparison of details, which can ultimately be rendered proportional, is due to the fact that mental images are aggregates of mental impressions. These impressions are the qualitative elements in properties. In fact, it is the recognition that there could be a proportional comparison of the properties in one image as compared to those of another which makes possible the conversion of these images into concepts, by means of which the comparison can then be made.

For, while concepts are always supported by images, the images alone lack the definitional rigor of a concept. So it is only concepts which lend themselves to explicit considerations of proportion. Between concepts, two or more properties may therefore be made quantitatively compatible, which means that they are commensurate and a precise comparison can be made.

An example of this would be a comparison between two shades of blue. Initially, both shades are seen as alternate images of one color. Then an enquiry is made as to how they might differ. It is asked, what is it which makes one distinct from the other? This yields two concepts of blue. For it is determined that the one blue is a lighter shade than the other by such-and-such a degree.

At times the degrees may not be determined internally. It may be that they are determined by context, say a broader range of shades, either in the color under consideration or in another color by comparison. In other words, it might be asked, what would this shade be in any color? For it may be decided that, like other colors, blue has both dark and light pigments. As the balance of dark and light reaches a certain crescendo, generally a mean point in moving from dark to light, the color becomes more pronounced. It is more clearly discerned. This is what an artist refers to as chromatic purity, or saturation.

As the blend of pigments moves further up the scale toward lightness and away from the mean, the color begins to decline.

The Limits of Reason

It fades. Qualities of light and dark may not actually be counted when these determinations are made. Nevertheless, for the purpose of making a distinction, it is assumed that a precise incremental difference between the various shades of blue does exist. It is a matter of arranging the various shades on a graduated scale. Or, at least, it is a matter of assuming they could be so arranged. Thus, in the relationship between either the qualities or the properties of objects, proportion is being determined.

But the qualities or properties may also be incompatible and therefore incommensurate with one another. This is tantamount to disproportion, since proportion cannot be determined. It is demonstrated in the incompatibility of hue between red and blue, a disassociation which can only be remedied by a relative comparison of wavelengths.

But from the perspective of this philosophy, wavelengths are phenomena which are only theoretically correlative to color. They are not intrinsic to it. In the grand design of experience, correlated phenomena consistently appear in causal relation to one another. However, as the connection cannot be affirmed beyond its consistent correlation, the correlated phenomena must be viewed as independent of one another.

It should be noted here that proportion, commensurateness, and incommensurateness are terms which are generally applied to relationships between extensions. So a dissociation can readily be demonstrated in the incompatibility which occurs between some extensions, such as between the concept of a Euclidean straight line radius and that of the continuous, unvarying curve (circumference) of a perfect circle which is derived from it.

In this case, a comparative proportion between them simply cannot be registered. For, in accordance with Euclid's definition, there must be an indeterminate number of equal radii falling upon the circumference to assure that it is an unvarying curve.

> A circle is a plane figure contained by one line such that all the straight lines falling upon it from one point among those lying within the figure are equal to one another.[36]

Anything short of this indeterminate number of equal radii would allow the possibility of a change in the curvature of the circumference. So, as a result of the need for an indeterminate number of radii, the ratio between the circumference and radius can have no determinate measure.

By the same definition, if the measure of the circumference is indeterminate, then the measure of the radius must be determinate, or vice versa, since the uniform length of the radii ("all the straight lines…are equal to one another") is what limits the circumference to an unvarying curvature.

Thus, if the radius is determinate in extension, the circumference is not, though it appears to be. For it is merely figuratively drawn to appear so. Accordingly, there is no comparison between the lengths of the radius and the circumference, no rational means of associating the one with the other. The radius and circumference are therefore incompatible. Consequently, they are incommensurate in relationship to one another.

But, to return once again to the general topic of proportionality, when concepts of objects are compatible by definition, they are commensurate. Thus, to employ a simple example, a con-

---

[36] *Euclid's Elements*, Book I, Definition 15.

The Limits of Reason

ceptualized line with terminating ends which is composed of ten units is compatible with such a line composed of twelve units.

The properties of objects can be conceived to occur in aggregates: a certain amount of this or that. So the role of aggregates of properties is as follows. Multiple qualities are associated into properties, such as a stone having a property of being composed of a specific blend of minerals. The individual minerals give it its qualities of hardness, color, etc.

The latter can be individually designated as either qualities or properties. Or they can be combined as a single property, such as the property of a stone being an emerald. So, a single quality can be considered as a property—i.e., greenness. Or the stone's being an emerald can be its salient property. In this case, the color green is one of the qualities giving the stone its property of being an emerald. Thus it is properties which are generally compared as aggregates.

A consideration which has been brought forward in this essay is that of the feasibility or infeasibility of a precise association of properties between objects. Aggregates of properties are thus compared between objects by means of a proportional relationship. Accordingly, a person can assert that a specific stone, as compared to another, has more or less of a particular blend of minerals, the blend being a property and the amount of it being an aggregate. Or he can point to the quality of color alone as a property and maintain that an emerald in his possession is greener than one which is not in his possession.

An extension, which is the most fundamental characteristic of a physical object, is an association of all its properties considered independently of and expressing spatial contiguity with other objects. Thus both its independence of and its position relative to another extension is an expression of its properties.

A stone is an extension. The shape of it is a property. The shape may be considered the result of certain qualities. The shape, in turn, determines the stone's position in relationship to another contiguous stone or to the ground upon which it lies. And the contiguous stone and ground do likewise in relation to it.

As another example, the length and flexibility of a piece of rope are properties. Its length expresses extension. And its flexibility, altering its figure, determines a variety of positions. In each of these cases—the stone, another stone, the ground, and the rope—the properties are physical. These objects are physically extended. But, in contrast to them, it can be observed that, since a Euclidean line is ideal, it has no physical properties.

A Euclidean line is nevertheless extended. It has the nonphysical property of being extended in one dimension only. Yet, like a physical extension, the line may be understood to be composed of units, or parts. The units are, of course, nonphysical like the line. Nonetheless, an aggregate number of them may be established, based on the length of the line and the selected length of each unit.

However, the aggregate of such units of length in a radius must remain imprecise in its ratio relationship to the circumference of a perfect circle. It is so when the radius is determined to be a rational length and, as a result, some multiple of pi becomes the length of the circumference. It must accordingly be asked, how many units are there in an irrational number? So the aggregate of units of the circumference cannot be compared to the aggregate of parts of a rational radius which is a straight line terminating at both ends.

Thus it follows from these observations and others like them—such as an examination of Euclid's definitions of a point

The Limits of Reason

and a plane surface—that the entire thirteen books of the *Elements* constitutes a theory built upon incommensurables. The definitions are not only incommensurate with nature. They are incommensurate between themselves.

But the fact that this is not mentioned in Euclid is not surprising. It would be like informing a driver that the bridge he is crossing is resting on air. Such information about Euclid's geometry could have resulted in a neglect of his powerfully effective system. And it would have undermined Greek geometry in general. But this was not done in ancient times. Nor were the principles of Euclid's geometry seriously challenged until the nineteenth century. For it is the practical effectiveness of a system that truly matters.

Commensurables, in the extended application in which they are employed in this essay, are important in human thought. For relations between conceptual classifications of any kind are either commensurate, or there is a suggestion that they are (even when they are not). It is a matter of one concept being related to another according to properties which are common, or presumably common, to them both. So, if this should not be definitively the case, it is a matter of it being made to appear as if it is. For human beings cannot think rationally without these relations between classifications.

But, regardless of the importance of commensurables in rational thought, it is incommensurables which are in fact the reality of human experience. For this reason, the concepts brought into a commensurate relationship to one another in rational thought are in fact idealizations which may often not have commensurate properties in common.

Since even the most ordinary concepts are abstractions from experience, they must be seen as products of human invention, however closely they may seem to adhere to experience. Thus, when understood subjectively as thought, conceptual objects take on a different reality. They obey rules of reason which are not found in physical experience.

For rational thought to take place, a logical correlation must occur between any two concepts involved in a progressive chain of reason. But it may be that such a correlation does not pertain. So it must be assumed. In other words, the logical correlation may be based on properties which are contained within each concept and which can be associated with one another. Or it may not.

Accordingly, a logically-based assumption may be as commensurate with another in terms of properties as "all dogs are mammals," in which dogs and mammals have kindred properties, such as warm-bloodedness and lactation. Or the assumptions may involve concepts as incommensurate as the relationship of a radius to the circumference of a circle.

Hence the fact that, in spite of its complexity, the introduction of a perfect circle into Euclid's geometry takes the form of a definition, rather than a proposition. As a definition, the established relationship between the radius and the circumference, however questionable, is not to be disputed. For, if it were, as would be the case in a derived theorem, or proposition, the lack of a common measure between these two properties would be evident.

In human awareness, the process of abstraction into thought rarifies a concept and sets it alone in an ideal realm of its own. That realm is the mental realm. However, the degree of idealization can vary greatly. It may range from a concept which closely follows experience, omitting but few properties, such as

The Limits of Reason

the concept of a horse, to a concept which is a product of imagination, such as that of a unicorn.

Now thought is shared by imagination and reason. This is to say that thoughts are composed both of images and concepts. Thus the statement "a dog runs through a field" can take the form of a series of images, until the need occurs for the properties of one image to be considered specifically in terms of their relations to those of another. Then it becomes a relationship of concepts.

For example, a person may call to mind images of a dog running through a field. It is in this manner that he may imagine the situation. Such a train of images is concrete. This is to say that it more or less directly correlates with experience. Thus the thought remains largely faithful to the mental impressions forming the experiences of a dog, a field, and a running motion. It is "sensual" in its fidelity to the various mental impressions received. At least, it is so insofar as the rough texture of a person's mental impressions and image-forming faculty of imagination will allow.

But, if he should go on to think, "dogs are mammals," he has begun to conceive dogs as a classification, and to compare this classification with another classification, mammals. So he is defining the concepts, "dogs" and "mammals," and relating the properties common to them both. For it is specifically the properties of concepts which are pointed up by their definitions.

In doing this, he draws upon memory.[37] For he, or someone else, has at some past time abstracted certain properties from experience and created definitions from an observed correlation

---

[37] The peculiar character of memory in an immaterialist philosophy is discussed in the book, *The Immaterial Structure of Human Experience*.

of properties within an object, range of phenomena, or an event. In doing so, he or that person has formed higher abstractions from the mental impressions combined in imagery. And he has done this for the purpose of developing associatively related concepts. Concepts allow him to think rationally by means of a logical train of thought based on the associations between them, or a supposition of such associations.

The same sort of thing as has been discussed concerning these general concepts is done with the more specialized geometrical figures or arithmetical numbers. If a person should simply think "circle" as an image without defining it, he can find many examples of it in nature: a cartwheel, a coin, the sun, a full moon, etc.

But if he defines a "perfect circle" as Euclid does, he has abstracted from the image or images and formed an independent concept, the principal characteristic of which is its capacity for forming a relationship with other concepts. However, it does not need to correlate—i.e., share properties—with those other concepts. It only needs to appear to do so.

Thus, if it is held that Socrates lived in a modest house, that all people who live in modest houses are philosophers, and none who do not live in modest houses are philosophers, therefore, Socrates was a philosopher, it may be asked, what is it about modest houses that makes the people who live in them into philosophers? A person would be hard put to make such a determination.

For any argument about people living in modest houses because they are contemplative and do not waste their time and resources on conspicuous consumption could certainly not be drawn directly from the character of a modest house. Nor could it be definitively drawn from the character of a philosopher.

The Limits of Reason

There are no identical properties in the two concepts, modest houses and philosophers, which can create a definitive association between them. Rather, they are connected by an assumed motivation and the plausible expression of that motivation. Even if the above statements appear to be grounded in logical form, it is this assumption which is sealed with the universal quantifiers "all" and "none."

Granted that most well thought out arguments are more carefully based on observation than this example. Nevertheless, the fact remains that leaks do eventually appear in many logically secure chains of thought. Hidden assumptions become less convincing over time and are eventually challenged. This can occur because a direct correlation did not exist between the concepts in question.

A classic example of this is expressed in Richard Dedekind's essay, "Continuity and Irrational Numbers."[38] Here he develops his argument upon an assumed parallel between number and a point on a line. But the assumption implies a relationship which does not hold. For Euclid defines a point as "that which has no part."[39]

Whereas, as discussed in the present work, any number is expansive. It bleeds into the numbers nearest in proximity to it. This is because, however variable it is in magnitude, it is an imaginary extension. For it is composed of arithmetical units which are extensions. So numbers exhibit the property of extension. And no matter how minute the extension may be, it is still

---

[38] Richard Dedekind, "Continuity and Irrational Numbers" in *Essays on the Theory of Numbers*.
[39] *Euclid's Elements*, Book I, Definition 1.

an extension. Thus the points on a line, which are by definition not extended, cannot be held to be equivalent to a number.

Now, as has been previously indicated, arithmetical units are the foundation of numbers. They are idealized thoughts because they are stripped of all physical properties except extension. And they are defined as being identical to one another, which identity is also ideal. As a result of this idealization, arithmetical units compose numbers which are referenced to a mentally conceived number line, rather than directly to the physical realm.

It is in this way that they become mental abstractions without any specific reference to particular physical entities, until so applied. They are applied to physical entities for the purpose of quantitatively organizing them for thought. And, when this occurs, the physical entities are themselves generally conceptualized for thought.

However, conceptualization is classification. And classifications cannot be found in nature. It is multiples of qualities which are perceived within the mind. And they are often encountered in association with one another. That is, the grouping of qualities into properties and the recognition of individual properties as belonging to specific objects is based on the sequences in which the mind receives its impressions, particularly if these are repeated sequences.

This collation of qualities and properties into associations is recognized as forming extensions. And extensions are objects, be they physical or mental objects. But an extension apprehended in terms of a mental image without the aid of conceptualization remains vague. While images of perceptions may be quite close to physical experience, they are not exact and may shift in their details while being held in the mind. For an image is undefined.

The Limits of Reason

Classifications, on the other hand, which are derived from images, are made precise by definition. Thus they are concepts. And nothing directly correlating to these can be found in physical experience. For they are products of the mind. That is to say, even those concepts derived from what are understood to be sensory-based images are yet remote from physical experience. In other words, a concept is set apart from the direct imagery of experience by its submission to a definition. For a definition points up certain properties and omits others from consideration.

Classifications are applied to physical qualities which are initially associated together in the mind as perceptual images. The perceptual images are relatively faithful to their physical origin, give or take some qualities which may be overlooked. In other words, the repeated associations of qualities may not be exact duplicates of one another, leading to the omission of some qualities in the general image which is held within the mind.

The subsequent conversion of the image into a classification gives the intellect an orderly means of interacting with objective experience. For it is by means of their stability and consequent receptiveness to reliable associations being made between them, that concepts lend themselves to orderly thought. But this conversion into classifications is accomplished at the price of a further reduction of qualities, as the assignment of a definition to an image places an emphasis on certain properties at the expense of others.

It is in this way that classifications are created to make conceptual sense of perceptions. But the concepts thus formed clearly delimit the physical extensions they represent. This places them remotely in the mind and binds the original images (from which they are derived) not so much to experience as to the rational processes of the mind.

Once physical images, like dogs or trees, are converted into general concepts, numbers may be applied to multiples of the concepts. This is because defined concepts delimit the mental images of physical extensions: they are precisely such-and-such and not something else. So the conceptualized extensions, having been derived from physical experience, are now susceptible to enumeration. For they have a closer affinity with numbers.

The immediate reference of numbers is to the number line rather than to physical experience. So, due to the greater level of abstraction in the realm of numbers and its complete removal (aside from the property of proportion) from physical experience, numbers can be used to make precise quantitative sense of multiple conceptual abstractions representing physical objects. For neither the numbers nor the conceptual abstractions of objects participate directly in physical existence. Of course, numbers may also be applied to images. But such an application is of little use in structured thinking.

For example, if a person should encounter five apples in a bowl, he might count them and say, "There are five." But what does he mean? Five of what? Since each apple is different from the others, it does not exist in a quantity of more than one. So it is only in terms of a generalized concept of apples that there are five. This is true, even if it is simultaneously acknowledged that there are differences between the apples. For, in counting, the differences are ignored.

Since numbers are abstractions, they are applied to other abstractions derived from such disparate objects as stones, trees, dogs, and the like. These objects have a physical origin. But some portion of the particularity of the object must be held tem-

The Limits of Reason

porarily in abeyance in the mind, if the object is to be counted. Thus the thought "four trees" will ignore what makes an individual tree unique. Rather, what is thought about is enumeration in terms of a general reference to the thing enumerated.

It is for this reason that a generalized notion of a tree can be carried along in the thought process in a manner like that of an algebraic variable x, which perhaps represents the number 4 in an integrated sequence of algebraic expressions, where the x appears in various combinations with other values. The tree can be treated in this manner in the ongoing enumerative thought process. Until the thought process has been completed, it is as though it were undefined.

Concepts have the advantage of various degrees of generalization. Pine trees resemble fir and spruce trees, and may be imagined to do so in more or less degree. But, conceptually, if trees in general are under consideration, only those properties which define any tree are relevant. All conceptual abstractions, however inclusive or exclusive their encompassment of properties, are precise and discrete in this way.

It is this discreteness which makes a concept commensurate with itself when employed in different contexts. This is what gives the separate usages identity. So a pine tree in one context is a pine tree in another. And, in a broader sense, it is this discreteness which makes one concept more loosely commensurate with a different concept. Thus dogs and mammals, which are different in their encompassment of properties, are nonetheless commensurate to a degree. For the stability of their definitions provides a reliable comparison between those properties which are relevant to both.

So a generic definition of a tree (indicating any kind of tree) makes a pine tree definitionally identical to a fir or a spruce in spite of differences in shape, cone position, needle length, and bark. This is because these properties are left in a state of abeyance by the definition. Consequently, they do not enter into the comparison. Likewise, one perfect circle is also definitionally identical to another because size is excluded from the definition.

One number 4 is identical to another number 4 because these numbers are composed of a like quantity of identical units, which units have been rendered devoid of any properties other than extension and a variant magnitude. Whereas a non-mathematical commensurability, such as the relationship between dog and mammal, is looser in application than a geometric or numerical commensurability. For commensurateness in mathematics is more narrowly defined than it is for concepts reflecting physical experience. This is because fewer properties are involved in the former.

Concrete circles, which are less uniform with one another than perfect circles, are widely found in cart wheels, pumpkins, and sun dials. Numbers, on the other hand, cannot be located in nature at all. What is worse, they bleed off within themselves. It is this bleeding characteristic which is most troublesome about numbers. For, as they are pure abstractions, a person might think that this ought not to be the case. But an expanded number line's progressive figures can be seen to blend into one another.

For example, a number like 4.999… is not considered to be irrational, since the decimal is repeating. But neither does it appear to be fully rational, since it is not discrete, like the number 5. Rather, like an irrational number, it is indeterminate. It approaches the number 5, but is not that number. Neither is it the number 4. It fills the gap between 4 and 5, bleeding towards the 5, but not entirely.

The Limits of Reason

It is this kind of indeterminacy which characterizes the physical world as well. Thus everything shades off into something else. This is the bane of the human mind, which desires order, but which allows for a recognition of some indication of disorder because of the complexity of every kind of relation.

Since conceptualization is a mental process, and mental processes are quite flexible, there can always be controversies over how distinctions should be made. Thus it can be said that, prior to its intellectualization, the physical world is no more than a loose pattern, but a pattern nevertheless. It is a pattern because mental impressions are received consecutively, and often in association.

But it is a pattern which falls short of a necessary discreteness, determinateness, and commensurability. On the other hand, the realm of human thought and intellectual awareness is increasingly discrete, determinate, and commensurate, the more mental impressions are converted into concepts, and the more those concepts are refined.

But are not the conceptual classifications themselves somehow embedded in nature? Cannot a gray rock exist separately from a brown rock and a quartz rock from sandstone or shale? And is this not in some sense immediately evident in terms of initial impressions on the mind, generally termed sensory?

Of course, associations of properties can be recognized prior to any deliberate intellectualization of them as classifications. But the fact that several rocks, differentiated yet lying together, can be encountered as separate extensions does not belie the role the human mind plays in forming these distinctions.

The classifying human mind is involved in synthesizing experience both at the physical and at the conceptual levels. What is important is that it is not at the physical, but at the conceptual level, that the mind is able to employ classifications as a function of their definitions. At the physical level, the delineation of extensions and their qualities occurs as though it were immediate. Yet it is the classifying faculty, applied to this more fluidly imaginative state, which fully articulates the delineation.

So an association of qualities is recognized when impressions first make their appearance in the mind. But it occurs imaginatively without definition, without a clear recognition of qualities grouped as properties, and thus without the consequent rigor which would follow from such a firm articulation.

Consequently, the delineation remains tentative until an acknowledged act of classification does occur. That is why the associated impressions on the mind are initially worked up into an image. The image is a bounded extension. But it is free to shift somewhat in its qualities, and thus in the precise delineation of its properties, so long as those properties are not deliberatively set by a definition.

These conceptual classifications of the deliberative mind differ from the more flexible groupings of mental impressions into an image. But they do not differ entirely. And it is this correspondence between the conceptual and imaginative operations of the mind which constitutes both the power and limitation of human material awareness.

For, as to power, the flexibility of imagination is creative. And the rigor of reason (conceptual thinking) organizes experience, subordinating it in whatever degree possible to the human will. But, as to limitation, there is the problem that beyond the imaginative and conceptual functions of mind lies an undifferentiated whole, the incommensurate realm of disordered experience.

The Limits of Reason

The imagination's projection of images representing extensions (i.e., physical objects) may be endlessly varied. For example, the initial perception of an apple may include its stem. Subsequently, it can be seen that the process of converting this experience into a member of a uniform body of knowledge composed of concepts with precise definitions will involve pretensions of finality. For invariable relations will be formed, as in the determination that an apple is the fruit, not its stem. But those relations are products of the mind. So they cannot be held to constitute an incontrovertible claim to truth.

Impressions on the mind begin the process. Thought, employing both imagination and conceptualization, tentatively completes it. But only tentatively. For it is a procedure applied to a domain in which a rational structure was not originally apparent. In other words, the human mind is not spontaneously aware of the objective world in a final state.

Rather, it perceives a world which is bewildering and disordered in its initial character, about which, nevertheless, certain primitive organizational structures have already been reached in infantile and early childhood development. These are modified, expanded, and built upon with experience and with an absorption of preexisting formulations of that experience, which are learned.

Thus the contribution of reason is ideal. It is more faithful to itself than to the imaginative material constructed from the mental impressions which it encounters. From this it may be deduced that those concepts which make up the terms of every proposition have concealed within them the unreality of the ideal. Conceptual systems, therefore, can only make a sophisticated map. They cannot produce an indelible representation of an objective world.

But let both the imaginative and conceptual apprehension of experience be grouped together and considered as one integrated process. Let it be said that collectively they constitute the making of a structure which is accepted as the realm of material experience. That does not fully explain the human mental relationship to its experience.

For the mind is aware of its own consciousness. And, external to its material awareness, there must be some explanation not only for the various mental impressions received, but for the order in which they are received. Otherwise a solipsism has been committed. Human awareness floats free of any transcendent or external source. It is enclosed within itself.

It must be conceded that prior to the work of the mind, experience is ungrouped and unclassified. This is to say that without organization it cannot be imagined or understood. But the problem which remains is that, since the mental impressions and their order of reception in the mind are independent of human knowledge, these impressions would appear to be ontologically antecedent to the imaginative and conceptual matter which falls under the operations of the human mind. In other words, the origin of this experience is estranged from human awareness and understanding.

This is a problem encountered in Immanuel Kant's *Critique of Pure Reason*. For, unlike the immaterialist view expressed in the companion book to this work, *The Immaterial Structure of Human Experience*, he proposes an unknown thing-in-itself. Thus his phenomenal realm appears to float disconnected above a completely inaccessible noumenal realm. That is, inaccessible in terms of the order of presentation of his manifold of sensation, other than whatever order may be imposed upon it by his

The Limits of Reason

intuitions of space and time. The phenomenal realm is without a foundation. Its categories cannot be traced to a source.

That is an unacceptable situation. For it leaves the human mind intellectually bewildered as to source. Some degree of mystery in the inaccessible workings of spirit, the source of mental impressions, may be conceded. But not a complete void in human understanding of the relationship between experience and its ground.

So it must be said that those deepest relations of phenomena which lie outside the human phenomenal realm, beyond the reach of imaginative and conceptual awareness, are nevertheless hidden in the dynamic of spirit, or universal consciousness. Whatever their relations, they are embedded in this greater domain.

And these deepest relations express a correlation to what is known to the human mind. They are not completely estranged from it. For human consciousness is a direct, limited expression of universal consciousness. So, between the human mind and the source of its limited experience, a working connection exists.

Those deepest relations of spirit must not only transcend the associative and classifying operations of mind in both its imaginative and intellectual operations. They must be assumed to be beyond human comprehension in their complexity. That is, potential relations at the noumenal level cannot be made actual to the human mind. The human mind cannot understand how spirit works within itself.

Rather, those relations are mediated by spirit, which in the individual person is focused to a particular human perspective. This focus, or self-limitation of spirit, supplies impressions to the mind in particular sequences and associations. Human intelligence develops the impressions into the structure of material experience.

This includes both the mental structure of thought and the spatial structure of physical experience, which in the final reduction is also thought. For what are interpreted as physical objects are images in the mind. Human intelligence also works conceptually to interpret experience, converting multiple images into classificatory systems. Consequently, the mind is limited in its use of materials. For both its physical and mental experience are constructed from the mental impressions.

Given this explanation, any associated mental impressions, which are interpreted as physical and thought extensions (physical objects and objects of thought), must be assumed to have an origin in spirit. In turn, any images or concepts formed by human imagination and intelligence are assumed to derive their origin from these impressions, as well as from the order of reception of these impressions in the mind. Images, and especially concepts, may be more or less faithfully representative of the mental impressions, depending upon the degree of simplification of the image or idealization of a particular concept.

Nevertheless, though the mental impressions and their initial order of presentation to the mind are derived from universal spirit, they may not be imposed as limitations upon it. In other words, human material experience may be valid insofar as it serves the finite character of human beings. But it is not binding on that universal spirit from which it sprang.

Furthermore, it should be understood that any associations or classifications which the mind makes of its mental impressions will not exhaust the possible range of associations and classifications. These can change. For there are yet other relations in spirit which remain as yet unrealized in human experience. The-

The Limits of Reason

se unrealized relations of spirit constitute the multitudinous range of the possible as opposed to the actual.

They are the reason a future lies open to human awareness, while the past appears to be locked in rigid conformity with itself. For the future is inaccessible to the human mind, except where a uniformity of appearance, such as is identified in causal relations, may be expected or hoped for. The consistency of these causal relations is determined by spirit.

George Lowell Tollefson

"Infinite" Sets

It was demonstrated by the mathematician Georg Cantor that the series of all natural numbers and the series of all even natural numbers may be paired and extended to infinity.[40]

$$1 \quad 2 \quad 3 \quad 4 \quad 5 \quad 6 \quad \rightarrow$$
$$2 \quad 4 \quad 6 \quad 8 \quad 10 \quad 12 \quad \rightarrow$$

This reference to infinity should be a reference to indeterminacy. For the finite parts of a whole cannot create an infinite whole, since that which is infinite is *not* finite in any sense, as indicated by the prefix to the word "*in*finite". But let this objection be put aside.

What is of concern is that this denumerability, or pairing of the two series, is presumably accomplished by omitting considerations of quantification—i.e., by omitting an acknowledgment of the ordinal character of the set of numbers in each series. But is this so? It would appear that quantification remains very much in evidence. For the row of even numbers can be generated by doubling each of the numbers in the first row.

Even if this observation were to be ignored and the issue of doubling should be relegated to a merely descriptive purpose, the problem of quantification could not be removed altogether from consideration. This is because arithmetic is a science built

---

[40] Bertrand Russell, "Mathematics and the Metaphysicians" in *The World of Mathematics*, p. 1583.

The Limits of Reason

upon the unit. All rational numbers are multiples of units. And all irrational numbers are approximations of the same.

Thus a reference to numbers, or even to a numerically unspecified set of entities, is a reference to multiples, which turns out to be a reference to numbers. A reference to a numerically unspecified set of entities would, of course, exhibit their cardinal character, as in aleph-null representing a denumerable transfinite set. But such a character would be illusive. For transfinite entities, like the lists of numbers above, are said to be infinite. And any set of entities, when it is extended to infinity, is extended by count: by the adding of additional increments.

There is at least a repeated count of 1. To add is to increase by a certain amount. And to uniformly increase is to move quantitatively in some form of incremental manner. This is important because each row is assumed to be infinite. And the infinity is established by one more member being continually added to whatever number one wishes momentarily to consider to be last in the sequence. This involves consideration of an ordinal progression. Thus, to determine an infinity of the series, an ordinal progression is introduced.

If any series of entities is to have meaning as a transfinite set, it must at least submit to a hypothetical count. Only in this way will its infinity be established. Otherwise, adding extra entities would be meaningless. In other words, what does it mean "to add" if not to increase the count?

If in any series of entities there were no such ordinal progression, such as a count, its progression must be called into question. For it could not be known what the members of the set do when extended out to infinity. Why should they be held to do so? More to the point, it could not be known whether they do or not, in fact, extend out to infinity. For the extension to infinity is predicated upon a rule: a rule of continual incremental increase. Where there is no such rule governing the progression, there is no question of an infinite extension.

Even the decimal digits of an irrational number exhibit such rules. One is that they cannot be random. For they are determined by their origin, as pi is determined by the relationship between the radius and circumference of a circle. In addition, there is the rule that, though this is an exact relationship, it has no common measure.

So the precise quantity of the irrational number cannot be determined. This is how the extension into infinity of its series of decimal digits occurs. Between this extension into so-called infinity and the fact that an irrational number cannot unfold in a random manner lies the basis of its being both irrational and a number.

Given these explanations, the following can be asserted. If there is anything meaningful in designating a set as infinite, then, concerning any infinitely paired series of numbers, or anything else so paired and infinitely extended, it should be seen that the extension of the sets is predicated upon a number line.

Therefore, so must the pairing be predicated on the same number line. For it is this ordinal relationship which articulates the one-on-one pairing between the two series. In other words, it is not just a matching of individual members, but a pairing of extensions (i.e., the two sets) whose infinite character is understood in ordinal terms.

Whatever the progression of each series might be, the two series are related by a count which is extended indefinitely according to rules determining a uniform progression. Numbers are a uniform progression. But where there are not numbers, there is nonetheless a number of something.

The Limits of Reason

Arithmetical science, which includes set theory, has the prerogative of temporarily abstaining from considering items in their ordinal progression. It can do this for the sake of performing various extended operations, or for the sake of a discussion of its terms in a meta-arithmetical context.

But it does not possess this prerogative to the point that it can be inconsistent with itself. In any string of operations, and even in references to the science overall, it must be consistent in its internal relations. Those relations are universally quantitative. In other words, a reference to the ordinal character of arithmetical entities is always relevant, whether brought forward or not.

It may appear to enhance logical clarity in certain situations if, for such purposes, a choice is made to ignore ordinal references. But, where this is so, the references are only being placed in abeyance. They cannot be done away with. For they must come forth when being called for in making a quantitative distinction.

As multiples of units, numbers are inherently ordinal. They reference a number line, which is ordinal. Otherwise they would not submit to a count. They would not be numbers. They would be no more than arbitrary symbols, which is to say that they would be symbols without meaningful extension. Thus any sets of entities which must be understood in terms of such symbols would be without meaningful extension.

It is no support for the abstract pairing of series above, when each series is considered in no light but that of a cardinally designated set, and when an omission of the operation of counting is validated by insisting, as Bertrand Russell does, that "we cannot ... use counting to define numbers, because numbers are used in counting."[41] For the counting numbers do not simply define number. Nor are they defined by number. They *are* number.

---

[41] Bertrand Russell, *Introduction to Mathematical Philosophy,* Chapter II, p. 13.

The Limits of Reason

Science and Philosophy

There is an ongoing intellectual tension for humankind which lies between the undivided unity of consciousness and the fragmented material experience which makes up the content of that consciousness. Science must work within the disparate phenomena of the material but is always striving toward that underlying ground of unity. Thus it attempts to move from particularity and limitation to universality and an all-encompassing oneness.

A typical scientific system seeks to encompass all that lies within its purview. Thus physical science strives to comprehend material phenomena in terms of energy relations. Yet, when it runs into contradictions and anomalies, it breaks up into subfields like classical physics and quantum mechanics until the problem can be resolved.

But its goal remains: to simplify and to create one grand system. So, though at present it is far from doing this, modern physical science appears to be moving toward unification. For, if the material world could be understood entirely in terms of energy relations, it would achieve the simplest, most practical explanation conceivable.

But, if this is an attempt to attain a unity like that of consciousness, it is an illusion. For human concepts are drawn from the perspective of the material. The material realm is the finite content of consciousness, while consciousness itself is unbounded and indivisible and therefore *not* finite. Such a divergence between the two is unbridgeable. Thus an attempt to achieve what cannot be achieved results in conceptual confusion.

Physical science relies increasingly on mathematics. And mathematics is sometimes seen as expressing conceptual transcendence. For it is thought that its first principles, or axioms, are innate to the mind, thus extending beyond the material phenomena to which they are applied. But, in spite of its appearance of abstract self-containment, mathematics is based on the same recognition of proportions that physical observation must take an accounting of. And proportions imply finitude. So there are no eternal mathematical ideas. No conceptual bridge can be found which leads from the mind to whatever lies beyond.

It is equally the case that energy is a concept rather than a physical reality. For what is immediately given to observation is not energy, but change and the quantitative expression of change. Thus it can be seen that energy is simply a concept for change and its proportional relations.

Now it must be conceded that the complex relations underscored by the ideas of force, work, and energy are obvious and useful. But a theory which allows for this is unnecessary to a simple understanding of physical experience. What is needed is a willingness and ability to approach the mind's relationship to its content in a more fundamental way. In other words, how is it that human beings know things?

The object of physical science is to work with dynamic material systems. For it has an interest in prediction. But, when viewed in its simplest terms, change alone is found to express these material relations. This is made evident when thought is applied directly to the temporal unfolding of physical events.

Science does not observe change in the physical realm to understand its general character, but in order to control and effect

The Limits of Reason

that change. To do this, it must grasp the various individual instances of change. So it is not concerned with a broad generality. Instead, it converts the natural order into relations of force.

It translates the quantitatively disparate, but mutually proportional, relations of changing phenomena into a power capable of effecting each individual change. To do this, it must create a theoretical system suited to its purpose. Thus the concepts of energy, force, and work. And thus the grip these concepts have on the imagination. For the human mind cannot accomplish its purpose without them.

George Lowell Tollefson

Energy and Change

What is energy? Where energy is presumed to exist, it involves relationships between things in space which are observed in terms of time. That is, change is observed. Physical change is then explained with the concept of energy. But there are different kinds of changing relationships. So different kinds of energy are invoked: electrical, thermal, etc. Where change is actually observed to take place, it is considered a kinetic form of energy.

The problem with this procedure is that the mind begins to consider a term defining relationships between things as a thing in its own right. People begin to discuss energy as though it existed independently of change. In mechanics, they say that motion is a kinetic expression of energy. To account for its variations, they relate energy to mass. This allows them to develop a concept of potential energy, where there is no motion.

But, for greater clarity in illustration, let the phenomenon of light be examined. Light is considered a form of energy and as something which exists in its own right. But has light ever been observed as an object, rather than as a quality associated with objects? There is no denying the experimental effects in which what appears to be the behavior of light is observed. But, in spite of such convincing evidence, the nonexistence of the thing as an entity in itself must be insisted upon. For it should be asked, is light known independently of the phenomena in association with which it is observed?

Even if these be nothing but motes registering a beam of light, simultaneous reflections and transferals of images of objects seen behind and through glass, refraction in water, the

The Limits of Reason

anomalous behavior of the double-slit experiment, or what appears to be the coincident behavior of electromagnetic phenomena with light phenomena, all of these observations can be disassociated from theory and considered as isolated properties of extended objects. Though it cannot be denied that it pains the organizing mind of a human being to do so.

So let a moment be taken to consider the various phenomena of light as no more than simple expressions of change. A person enters a dark room and turns on a light, whereupon she observes assorted furnishings. The ability to see what she could not see, or could see only dimly, before is an expression of a relationship between herself and the things seen, not seen, or seen in a different way (perhaps dimly or in differing shades of color).

Either she could not see the furnishings in the room before she turned on the light. Or she could see them only dimly. She flipped a switch. A bright bulb appeared, accompanied by a coincidental sense of pressure on her eyes when looked at directly. And the furnishings of the room appeared clearly before her. Are these phenomena a product of the mysterious energy of light, which admittedly submits to a measure of proportionate changes in the totality of what is observed? Or are these phenomena independent changes in properties which happen to be proportionately coordinated?

In the case of shadows and colors, it is not light which is seen. What are seen are dark and light objects of various hues, which may vary in intensity or hue when approached. The light itself is simply a concept designed to account for the expression of change in the relationship between the experiencing person and the objects in question.

She observed these phenomena in one condition but a moment before: they appeared to be in deep shadow. Now she observes them in another condition: they stand vividly before her with subtle distinctions in their conformity of parts. It is in this way that there has been a change in their condition which appears to be the effect of energy.

If a dimmer switch is gradually activated toward increasing illumination, the changes will be subtle and seemingly integrated in effect. There will be an incremental brightening of shadows and heightening of colors. Or, if the light source is moved, these changes will occur in a different progression. Nevertheless, the changes are what they are. They are changes in properties, however fleeting in appearance.

It is because of the consistently coordinated measure of these effects that shadows and hues have been integrated into concepts which account for them in terms of the behavior of light—i.e., in terms of energy. Yet shadows and colors may be viewed simply as varying properties of the objects in which they adhere. Their degrees of brightness and saturation may be understood as changes in the qualities which make up the properties, i.e., as changes in impressions on the mind.

There is no attempt here to suggest that this perspective should be adopted by physical scientists. But there is a strong suggestion, even insistence, that theirs is an instrumental body of knowledge, not a final accounting of human experience. There is no question that it would not do for those who wish to calculate means for inducing effects in nature, that they should take the path of direct representational reasoning. But if the attempt is to understand experience in terms of its registry in the human mind, then that effort should begin with the mind. It should be asked, what is it that a person actually experiences?

No one experiences energy. What are experienced are sensations and changes in the sensations. Sensations are recognized

The Limits of Reason

as impressions on the mind. They are frequently registered in the mind in association with one another. And the origin of these associated mental impressions is attributed to physical extensions, or objects, in which the individual impressions are understood to be physical qualities. It is changes in the sequence and association of these qualities which are found to occur in predictable proportions.

So an empirically disposed person thinks: What can be done with this? She then systematically observes, measures, and labels the proportions of change in terms of a concept which she calls energy. But, following this, she will no longer be inclined to say simply that the proportional changes *are* energy. She will fall into the habit of saying they are brought about *by* energy. This in spite of the fact that she has never observed energy as an object extended in space and time.

George Lowell Tollefson

Space, Time, and Motion

Motion and distance are mutually exclusive concepts. Motion refers to both space and time, while distance refers only to space. Distance is a simple enough concept, since it indicates the amount of space between two locations. It has the same meaning insofar as it is a component of motion. So it is time which distinguishes the concept of motion from that of distance. But what is time?

When a person speaks of motion, he recognizes that it involves change—a change of location, which is a change of distance. But a change takes place in time. And time is recognized in terms of a comparison of changes, as in a comparison between the distance the second hand of a stopwatch moves and the distance a sprinter runs while the second hand moves. In other words, the time the sprinter takes to cover his distance is registered by the distance the second hand moves on the stopwatch.

But, if a sprinter covers a certain distance in terms of the distance a second hand on a stopwatch covers, then the movement of the second hand on the stopwatch must also take place in a frame of time which is determined by something other than itself or the distance covered by the sprinter. Otherwise, how would a person know that the stopwatch is registering time?

So, if time can only be recognized in terms of a comparison of changes, and changes can only take place in a span of time, a person seeking a definition beyond both change and time finds himself at an impasse. He cannot understand change without time. And he cannot understand time without change.

The Limits of Reason

Distance alone is a static spatial concept, since it does not involve any change. But the motion of the sprinter does involve change—i.e., a displacement in terms of distance. And the change takes place over time. Consequently, finding a definition for time remains a problem. It is certainly not as easy to define as distance. So what is it?

Distance is merely an extension, or a series of extensions, in space. Time, on the other hand, involves an extension in space being compared to another extension in space. A few millimeters of space is held in the mind in juxtaposition with fifty meters. That is how the distance the second hand moves on the stopwatch becomes a measure for the distance covered by a sprinter. It is his time.

What is important is that the concept of time breaks down into a comparison of changes in distance. But it results in an infinite regression, since each standard of measure must be set against some other. That is to say, the accuracy of the clock is measured against something else. That something else involves another comparison of distances. So on ad infinitum.

In addition to this, when taking all forms of change into consideration, the phenomenon of change reveals itself to be widely varied and often quite subtle, as in a chemical change and the freezing or thawing of water, as well as the larger-scaled type of motion already mentioned. All of these forms of change are motions, however subtle, which may be reduced to alterations in spatial location. Chemicals undergoing change restructure themselves, occupying microscopic amounts of space in different ways. Water expands or contracts, occupying more or less space. And the motion of the sprinter has been explained.

Each of these cases is submitted to a form of measure which is achieved by something that functions as a clock. That is, they are changes which occur against a background of change. In other words, any one change can best be said to involve a comparison of distances occurring simultaneously. But then the question of the simultaneity must be reckoned with. So, while time is intrinsic to an understanding of any change, it is itself a comparison of changes. And this is ultimately a matter of a comparison of distances.

The period of simultaneity is a span of time. But since any two changes, in which one functions as the measure of the other, must take place within a simultaneity of time, they in turn must be referred to a third change in distance demarking a separate measure of time. The separate measure determines the simultaneity. Hence the indefinite resolution of the problem of time. For time is an interminably regressive measure of itself.

This reduction of concepts to both time and change and nothing more basic than either of them can go no further. For these concepts both depend upon the same sequence of impressions which appears in the mind. So a question arises: how is it that human beings are confined to an experience of such sequences without knowing with any degree of certainty whence they originate? Assigning them to a source external to the mind leaves open what that source is, since they are only known in the mind.

So far, the best human beings have been able to do in their attempt to understand time is to observe changes in their awareness of spatial extensions and call the changes themselves the environment in which change occurs. One against the other makes time. But this is time defining itself. And it is obviously both circular and dead-end reasoning. Yet it is a problem science does not attempt to address.

To get a more subtle sense of the problem time presents, let an example be put forth. Let it be a bullet. A bullet is either

The Limits of Reason

moving or it occupies a specific location within a period of time. There is either motion or rest, which are the two physical states of inertia. In the latter state, the bullet being at rest means that it occupies a specific location in space for a specific period of time, however minute the duration.

When observed, this state of rest occurs in terms of a background. It occurs in conjunction with other states of rest, to which it may be compared. Or its state of rest, which is an absence of any motion affecting it, is noted in relation to a sequence of motions, or changes, occurring among other entities occupying space.

This is to say that the other changes occur in some uniform manner, such that a state of rest can be determined in relation to them. Moreover, a state of rest can also be observed in terms of a physical entity resting, or undergoing no change, in contrast to a sequence of changes unfolding among a person's thoughts, giving that person a sense of the stability and duration of the state of rest.

Or, if only imagined, the state of rest occurs in the mind against a background of images related to itself in one of the ways just mentioned. One or more of the elements of this background are the phenomena which are held up in comparison to the state of rest. These are either in a state of rest themselves or in a state of motion, or change.

Thus a state of rest, observed or imagined against a background of rest, exhibits no representation of time. For time is recognized by means of change, which involves motion, or spatial displacement. But here the object is at rest. And its background is also at rest. Consequently, there is no change, or motion, to register the effects of time. So the object's state of rest must be referenced to some other register of time: something changing somewhere which can be brought into comparison with it.

But the situation is different where other changes are immediately present. In such a situation, the object is at rest. But its background, or a part of its background, is in motion. These sequences of change, or motion, existing in the immediate background of the object at rest, denote time. So, in either case—rest against rest, or rest against change—it is to be noted that the state of rest is observed to occur within a time sequence, whose reference is conceived to be either external or internal to the person. For rest has duration.

This is so even though the immediate background appears to be changing. For that changing background is coordinated, or made uniform to the object at rest, in such a way that there is no change in its uniformity vis-à-vis the fact that the object is at rest. Say the background is in motion. Initially, it is moving at thirty meters per second, while accelerating at three meters per second, so that in each succeeding one second time interval it is moving at thirty-three, thirty-six, thirty-nine, etc. meters per second relative to the object at rest.

This uniformity allows the object at rest to remain positioned relative to such a moving background. A number of such elements in the background may be changing at different rates vis-à-vis one another. They need only be coordinated to the object at rest to act as determinates of its state of rest.

Thus an object can be determined to be in a state of rest in relation to a background which is changing. For it is the object alone which is determined not to be changing. Such a condition can only occur when there is a regularity of changes in the background which allows the object at rest to function as a stat-

The Limits of Reason

ic point of reference in relation to all the moving elements of its background.

The measure of the duration of this state of rest is dependent upon the particular phenomenal change a person may wish to employ in demarking it. It might be the sequence of his thoughts. It might be a rising or waning sound or changes in the weather. It might be a clock. Whatever it is, he knows that the usual standards of time mensuration—seconds, minutes, and hours—are no more than arbitrary measures chosen in accordance with an agreed upon convention.

What is important is that a bullet, or any other object, cannot be simultaneously in two inertial states. That is, it cannot be in motion and in a state of rest. It cannot be in the process of changing locations and at a specific location at the same time. For that would constitute a superposition of inertial states.

So should the flight of a bullet be squeezed down by differentiation to obtain an instantaneous velocity, location and momentum would appear to be expressed simultaneously. But this is not a physical condition. It is an imagined state of superposition. However, a person does not find himself compelled to think much about this at the macrophysical level because he is not confronted with the degree of precision which he is likely to encounter at the microphysical level.

So classical physics has long overlooked the underlying problem responsible for this dichotomy between motion and rest, which is that a moving object, such as a bullet, does not occupy any particular space. For it is always moving into or out of a particular space. An observation from direct experience should make this plain.

George Lowell Tollefson

Entropy

Given the fact that the human mind is conditioned by its awareness of the passage of time, what must be present to a human understanding is an order of existence which is progressively revealed but never summed up. For it is always becoming. Moreover, the passage of time appears to be without limit. This is the human experience of the material.

But, were time to be conceived as absent, the order of experience would be no longer progressive. And, lacking progression, there could be no causation as human beings understand it. Therefore, there could be no determinism. This view of material experience clearly has implications for the concept of entropy. For it may be asked: what is there that can be put forth in support of a concept which insists upon a combination of probability and determinism working together?

If a process unfolds, it creates a sense of time, and with it a possibility of causal relations. If it (and all things concerned with it) is in a condition of absolute stasis, then the sense of time is eliminated. The condition of stasis may originally have been caused. But now causation is removed from it.

In other words, what is suggested by the concept of entropy is a probabilistic process which unfolds in a deterministic manner and terminates in a final condition of indeterminism. The deterministic unfolding would imply that it is at least possible to know all the factors involved in the probability, and to know the relations of their unfolding. For, though it is true that these factors do not all present themselves to observation, the deterministic unfolding of the process reveals their concealed presence.

The Limits of Reason

To be more specific, take the accepted notion of heat transfer in a closed system. According to this idea, from which the second law of thermodynamics is derived, a statistically ordered dissipation of energy results in a future equilibrium which is mathematically determined. This implies that the probability in question unfolds in a deterministic manner.

Such a condition indicates that the probability, which takes the form of an ordered randomness, is not random at all. For it is certain in the unfolding of the events it describes. The apparent randomness is created by the inaccessibility of some of the factors involved in the process. Were they known, the situation would be determinate. And to hold that it is determinate is to say that it is caused. It is a temporal, caused situation which in its final state exhibits no time or causation.

This is the essential character of a closed system which evolves determinately toward a specific conclusion by means of an ordered randomness. In a finite system of this type—say where the molecules of certain gasses are being transferred to a condition of lesser excitation—it is not a problem. For the apparent randomness arises from a limiting condition affecting observation. The behavior of the individual molecules cannot be specified. But their collective effects can be. So their future behavior is collated probabilistically.

If the individual behavior of the molecules could be observed, this would not be necessary. For the velocity and trajectory of each molecule would be revealed. Thus the overall performance of all the molecules would be determinate. Consequently, the outcome is determined. And it is this determinacy which governs the process.

However, when the entire universe is conceived as a closed system, a contradiction arises. For the system of the universe, though evaluated probabilistically, is rendered determinate by its being a closed system. As a closed system, the infinite (indeterminate) universe is conceived to be also finite (determinate). Therefore, it is conjectured that all the energy within it, following certain observable patterns of behavior, will eventually dissipate to a state of full equilibrium.

Here an equivalence has been made. It is between, on the one hand, the unobservable individual activity of a finite number of tiny molecules, the behavior of which is nevertheless observable en masse, and, on the other hand, the unobservable character of every process of energy transfer in an unfathomably extended universe. The behavior of the molecules can at least be encompassed by general observation. Whereas only an infinitesimal portion of the universe can ever be surveyed by human means, or even by human imagination and intellect.

Yet, in spite of this disparity, it is assumed that the latter process can be evaluated in statistical terms, just as the former can. Thus a question arises: how can a determined outcome be statistically derived (much less derived in any other manner) in the presence of so much that must remain undetected and unknown?

Not only is an untold multitude of factors not available to observation. Any positive sense of the orderly continuation of the process under consideration is itself a matter of conjecture. In other words, a probability which originally served a practical purpose in limited circumstances has been converted into a final deterministic prediction of unlimited proportions.

Order

A fundamental law of the human mind is that there must be order. Without order the mind could neither perceive material experience nor form concepts about it. For chaos, which is the alternative of order, cannot be imagined. Chaos is merely represented to imagination as that which is not order.

But mental impressions must be arranged in some order to become an image. And it is from an image, or set of images, that a concept is formed. Thus, since there can be no precise image of chaos, the concept of chaos is that which is not imaginable, or that unrepresentable condition which is not order.

Order is proportion. It is proportion in the broadest sense, which is that in the human mind all things express a relation to one another. Since physical objects are spatial extensions, they are potentially quantitative in relation to one another by virtue of their being extended. Thus their relationships suggest ratios.

But, in specific cases, the actual relationship may or may not be commensurate. So it will not appear to be a ratio. Nonetheless, it does suggest commensuration. For the mind will make the necessary adjustment when needed. It will shave a little off of this and add it to that to bring the matter into a commensurate relationship.

So, if each spatial extension, when considered quantitatively, may be supposed to have ratio relationships with the extensions contiguous to it, give or take a few adjustments, then the totality of spatial extensions, when also considered quantitatively, will be conceived as mutually proportioned.

This condition holds for any structured view of physical phenomena, regardless of the immediate appearance of their relations. For example, it might have been thought that the ratio of a gas to a solid could not be determined, as they differ greatly in their characteristics. But at another level—the mass, the number of atoms or molecules, etc.—the human mind in its quest to establish order has rendered the relationship proportional.

It is in this way that a belief in a universal order has been arrived at. And this belief is what makes physical science, in fact all science, possible. It is also why so much of modern physical science has been reduced to mathematical relations. For quantitative order lies at the heart of physical science.

For an illustration of the utility of quantitative relations, take an arbitrary group of numbers, such as 1 6 5 3 2 4. No sense can be made of it as a number sequence, until an orderly relationship between the numbers is established. The simplest of these is 1 2 3 4 5 6. This is the beginning of the counting number line, which is an arithmetic progression. It is a progression of increments of one unit. These numbers are recognized as the natural numbers or, with the addition of 0, the whole numbers, which form a foundation for all arithmetical operations.

Now, if two ordered sets of numbers looked like 1 3 5 and 2 4 6, which are odd and even numbers, it would be apparent that 1 3 5 and 2 4 6 are each progressions by increments of two units. For the first sequence has been withdrawn from the natural, or counting, number line to create the second. In each of the new sequences, a different kind of progression from the counting numbers is established. But each of them can be referenced to the counting numbers because they exhibit a kindred order. Thus all three progressions can be referenced to one another.

The Limits of Reason

In this way, an unlimited proliferation of numerical relations can be made, which can be cross-referenced by means of the original order of the whole numbers. So, when these relations are applied to material experience, a sense of order in that experience is established. For not only are individual portions of material experience seen to be proportionate within themselves. The entirety of material experience is projected in the mind as proportionate.

Thus, if a person is restricting herself to the objective portion of material experience (what is generally referred to as physical reality), she can apply these various types of number progression and their interrelationships as means of indicating quantitative order in the physical universe. For they are all proportional to one another. In other words, they are conducive to simple or complex ratio relations. Consequently, many such means of establishing these relations are at her disposal.

So, by means of these various proportional relations, and by means of the fact that physical experience can be adapted to them, the physical universe is both assimilated and experienced by the human mind as exhibiting order. That is, it is experienced as fragmented into isolated pockets of order where the mathematics has been applied to it. These pockets of order are in turn proportionately related to one another.

But there is a limitation. The fragmentation of experience into pockets of order, as opposed to an immediate grasping of the entire physical realm as ordered, results from the fact that human experience is local and finite. Consequently, all things together can neither be experienced nor conceived at once.

Only local aggregations of order can be recognized. Over here is the sequence 1 3 5. And over there is the sequence 2 4 6. They are proportionately related to one another by their progression, which consistently adds two units. A greater order encompassing them both can be recognized in the counting numbers.

However, the human intellect continually strives for greater reach. It assumes that there must be a final all-encompassing order. For that would be universal order. Since order is the only thing which can be concretely conceived, this must be the case. Otherwise, human awareness would be enclosed in a rational box floating in an incomprehensible ether.

A person can only think both meaningfully and independently of, say, the numbers 1, 6, and 5 if she recalls a systematic ground supporting them. That ground would amount to an imposition of order. This need for order can be extended to any situation which might submit to a quantitative organization. Or any other organization, for that matter. Thus the expectation of order spreads like a ripple through the entire continuum of physical experience.

So, returning to the original two sets of numbers cited above, it can be imagined that 1 3 5 and 2 4 6 *should* combine into 1 2 3 4 5 6. This greater order would be feasible because the counting numbers have already been formulated and recognized. Thus it is clear that they must underlie the two smaller pockets of order. And the counting numbers will be found to underlie any greater range of quantitative order as well.[42]

---

[42] A person could not recognize order in any set of numbers, unless she first had a grasp of a system which embraced them all. That system is the whole numbers, which are made up of the counting numbers and 0. In one form or another, the first ten of these, 0–9, are expressed in any number. This would include negative integers, fractions, decimals, and even irrational numbers, since the latter are expressed in these ten whole number digits. Irrational numbers are in no way commensurate with rational numbers.

The Limits of Reason

Such an expectation of order is justified because it is the mind itself which creates order in the material realm. Mental impressions are received in a sequence. And they are found to lend themselves readily to associations. These associations form the physical extensions of experience, and thus the objects of thought as well. The physical extensions are further laid out by the mind in spatial relations which, though not necessarily being in a quantitative state of organization, lend themselves to it.

Thus, when the mind applies number to the physical, it recognizes what appears to a limited human perspective as a fragment of order. But it seeks to expand that fragment in expectation of discovering a greater whole. Three, six, or fifty stones, or some other manageable quantity of them, is all that can be encountered at once in physical experience. But, when pressing thought beyond immediate experience by means of a negation of the concrete, it is as possible to posit an indeterminate number of stones as it is to posit a universal order. However, neither can be formulated into a concrete mental image.

Thus they are not proportional to them. But the fact that they are expressed with whole number digits exhibits the human desire to bring them into as close a proximity to proportion as possible.

George Lowell Tollefson

Microphysical and Macrophysical

Arithmetic employs a systematic quantification of discrete units. That is, the discreet units of arithmetic form a logical system, the function of which is quantification. Geometry also works in a systematic way, using definitions, axioms, and theorems to construct figures. And in conjunction with arithmetic or algebra it is quantified. Once the algebraic form of these relations is in place and put to use in the physical world, mathematics becomes a language in which a conceptual trail need not be consistently adhered to. Results can be achieved by following the rules of the system.

So, in making use of this powerful tool for describing the physical world, a person must on occasion alert himself as to the nature of the mathematical concepts involved: discreteness, unit, quantification, and systematization according to axiom and theorem. How do these concepts relate to other kinds of concepts, such as the verbal concepts which are encountered in a description of the physical world: energy, force, motion, time, extension, etc.? How do they depart from them?

Modern physics has come to rely increasingly upon a practice of identifying its field of empirical investigation, then narrowing (without eliminating) this general field as near as possible to a purely mathematical structure. This structure is, above all, a systematic quantification of discrete units. In employing such a system, a person can proceed far down the road of computation without a close consideration of the descriptive concepts being dealt with.

The Limits of Reason

This procedure may continue until arriving at a quite distant point, at which it becomes necessary to return to the field of descriptive concepts. During the long interval, either in time or extent of the physical domain, these descriptive concepts have been allowed to be subsumed beneath mathematical operations. So the final transition back to descriptive concepts is not easily achieved, due to unsuspected conceptual complexities and contradictions which have arisen.

Admittedly, this is a special situation. For brief, practical computations are not what are being referred to. What is being referred to is the extensive acceptance and use of mathematical relations throughout a science, until they have become second nature to the mind of the scientist. This produces an inability to return to a broader descriptive field for two reasons. First, a close descriptive account of where the mathematics has been leading has not been taken fully into consideration. Secondly, unrecognized contradictions or difficulties expressed in the earlier descriptive concepts have only been brought into the open much later.[43]

For example, quantum physics has buried deep within its structure conceptual juxtapositions which would appear in classical physics to involve fundamental contradictions. One of these is complementarity, such as the idea that the position and momentum of a particle cannot be simultaneously determined, or the idea that particles are both particles and waves.

As to the former case, it has already been shown that the problem at the microphysical level concerning momentum and location also exists at the macrophysical level:

---

[43] For an interesting discussion of this problem, the reader is referred to Werner Heisenberg's *Physics and Philosophy*.

...a bullet, or any other object, cannot be simultaneously in two inertial states. That is, it cannot be in motion and in a state of rest. It cannot be in the process of changing locations and at a specific location at the same time.[44]

In the latter case, the wave concept describes an activity taking place over a geometrical space, while the particle concept focuses that same activity at a point or series of points within a geometrical space.

Unlike the wave, the particle, particularly if it is a fundamental particle like a photon or an electron, is considered in terms of point mass. This is to say that a particle has mass but is either unextended or is treated as being so. There are other such conceptual irregularities between the two levels of thought in physical science.

This modern complexity, which is found to be particularly troublesome in microphysics, resembles the old Ptolemaic system for describing the heavens. The older system was unnecessarily complex but worked quantitatively for some time, largely because scientific curiosity was not being systematically extended in that direction.

At the present time, a person would be inclined to say it was conceptually untenable. By this is meant that it was more complicated than it needed to be, those very complications growing out of descriptive irregularities and contradictions which lay behind the mathematical surface. Since those old conceptions about physical reality have been subsequently reworked through

---

[44] Quoted from the third paragraph from the end of the essay, "Space, Time, and Motion."

The Limits of Reason

Copernicus, Tycho Brahe, Kepler, Galileo, Newton, Lagrange, Laplace, and others, it has become possible to simplify the description needed to account for the movements of celestial bodies.

Likewise, concerning microphysics, it has become necessary once again to grapple with the descriptive manner of embracing reality. This project should be carried out first at the macrophysical level. Then, with a new macrophysical framework, it would be possible to return to the microphysical problem.

It should be remembered above all, that at any time in which a human thinker is forced to resort to probabilities in describing events, no matter how embedded they may appear to be in certain physical observations, as is presently the case in quantum mechanics, he is in effect admitting to, at best, an inadequacy in his data.

The inadequacy may appear to be physical and therefore irremediable. But in varying measure it can also be descriptive. Sometimes it may be necessary to accept such a discrepancy for awhile. For it can prove expedient in the short run. But this acceptance should be brought under scrutiny whenever circumstances will permit.

George Lowell Tollefson

The Conceptual Reduction

The problem with energy and force concepts, such as those employed in accounts of light, chemical change, physical change, and gravitation, is that they are no more than expressions of proportions in space and time. It is the time element alone which differentiates them from what might be considered purely static descriptions of matter.

In the case of experience of the physical realm, the issue is one of extensions of space. In the case of energy, it is an issue of change, active or potential, in the relative locations of extensions or of components within the extensions. In either situation the extensions are being modified over a span of time, which time is marked by the observance of one such change against others. Thus space and time lie at the ground of human thinking about energy relationships.

For example, a wavelength of light is held to be inversely proportionate to the frequency of light due to the fixed velocity of the light. In turn, that velocity is an expression of distance and time. Distance involves a set of contiguous physical extensions in space.[45] These form the background to any motion which covers the distance. This means that a differentiation between the extensions must exist in order to determine the distance. It is in this way that a certain number of entities, which

---

[45] These contiguous extensions include spaces occupied by vapors and gasses. Even a vacuum cannot be ascertained to be a completely empty space. For the act of determining its "emptiness" locates within it the means of making that determination.

The Limits of Reason

are the physical extensions, make up the background and measure of the distance.

Again, because the velocity of light is a constant, and distance and time are components of that velocity, the relationship between the distance and time is fixed. But they are in a proportionate relationship with the wavelength, which varies, and which is also understood in terms of distance and time. For a wavelength covers a certain distance at a rate proportionate to its length.

The rate is the frequency, which is a measure of the number of wavelengths passing a certain point. This is a sequence of recurring change, which is a measure of time. So, since distance is a measure of space, and since time is the other component of note, it is space and time which are the relevant concepts.

A wave occupies space in terms of its amplitude and the amount of space it is extended over in the direction of its propagation. So the wave is extended in both amplitude and length. Internal time with respect to the wave is its period of amplitude oscillation. The distance covered by a wave during its period of oscillation is its external displacement. That is its wavelength. Once the length of the wave is determined, the rate of exchange between one wavelength and the next is fixed by the velocity of light.

The particle theory of light also expresses energy in terms of space and time. But this is strictly external to the particle. For the particle itself defies internal definition in these terms. Because a photon is unextended, space is not expressed within it. Neither does the photon exhibit time within itself. For an expression of time requires an expression of space.

There can be no change within an unextended entity. For a change is an alteration in relations of distance. And, since this alteration in the relations of distance is the measure of time and the photon is not extended, neither does it express time within itself. It does not exist as a spatial or temporal entity. So, when speaking of a photon, a person is referring not to it, but to its effects.

The concept of an extended material entity undergoing change would appear to have an affinity with the concept of energy in that an object undergoing chemical or mechanical change expresses both space and time, which are the fundamental components in any expression of energy. For energy, which does its work in space and time, is the accepted means of accounting for that change.

But, considered strictly as a physical entity, the object is simply extended. By virtue of its extended character, it is not only an expression of space. Its integral parts are expressions of space. Only when the parts undergo change, or relative movement among themselves, do they express the internal time of the object. The parts can also become independent objects when considered as such.

The breaking down of a compound into its chemical elements can produce separate entities as well. But, as components of a compound, these elements are not parts of anything in human experience. For the nature of the object must change to reveal them. In other words, in each of these two cases, internal change involves some kind of movement. But the movement in chemical change is more complicated. For it involves a replacement: the substitution of several entities for another, or vice versa.

So it would appear that it is not the extensions which are understood in terms of time. It is the change involving their displacement which is understood in this way. This change is explained by means of the concept of energy. But change is predicated upon time: a change occurs over time. And time is predicated upon change: changes are the measure of time. Thus time is a nebulous concept, should it be considered independently of extensions and their relative displacement.

Space and time are also the reductive components of gravity. For a gravitational force is a proportional calculation of energy concerning one mass in relation to another at some specified distance between them. It is expressed as an inverse proportion because the force is understood to decrease as the distance between the masses increases. The force is understood to be evenly dissipated in all directions. So, as the distance increases, the force decreases in proportion to the expansion of the surface of a sphere.

When no motion, or change in the relative positions of the two masses, is being considered, the gravitational force is understood to be potential. But potential energy is an expression of anticipated change. That change is figured as an expenditure of kinetic energy in space and time. Thus energy is either potential motion, motion itself, or a combination of both.

This motion can be broken down into its fundamental components: space and time. When the motion of a body is being spoken of, it is as though the body were under the influence of a force—in this case, gravitational force. But its movement is not the gravitational force. It is motion. The concept of force is only being used to account for it.

Motion is understood in terms of space and time. So, if the effect of one gravitational mass upon another should be spoken of, what is being discussed is a change, or possible change, in spatial displacement. The components of this change are distance and time, or space and time, the time itself being a matter of relative changes in distance. This is what is experienced, not a force.

The Limits of Reason

Gravitational Force

Energy is a measure of change. Change manifests itself in the form of motion, or change of location. So, no matter the manner in which energy may be conceived, motion is spatial contiguity restructured by change. Something which was located somewhere in one moment will appear in a different location in the next. Thus the spatial contiguity of objects will appear to have been restructured relative to the object which is in motion.

In other words, the extensions of space will develop a different order vis-à-vis one another. The extension which changed place is one object among many others. When dropped, a spoon which was in the hand will have been relocated to the floor and will lie among the objects on the floor. Between those two locations, the hand and the floor, as the spoon falls, there is a stream of locations growing proportionately more distant from the place of origin, the hand.

This motion, with its accretion of variations in background, is difficult to represent to the mind as a single unity. So Isaac Newton appears to have initially conceived gravitational force without reference to any action. How else could he have imagined the inverse distance law of universal attraction between bodies? It is a law of proportions, not of action, though of course it can be used in calculations of motion.

But what does the word "attraction" mean? If it is thought of as an operative force, does it not appear to be a rather mysterious concept? What is the power which is the attractive force? And how is it that one attracted entity is brought under the influence of another? Perhaps what is implied in the concept of gravitational force is a convenient metaphorical reference, an illustration, rather than a description of something existent.

What is of interest is the relationship, not the cause of the relationship. Gravitational force functions to explain the orbital paths of heavenly bodies and the fall of a spoon. And the inverse square law describes the proportional changes involved in these motions. But it does not explain what gravitational force is.

A formula, $F \sim m_1 m_2 / d^2$, is given. It states that the gravitational force extending between two bodies is proportional to the distance between them. So, assuming the perspective of either one of the bodies, the extension of force from it to the other is equivalent to the surface of a sphere expanded away from a point lying at the center of the sphere's volume. That center is the center of the mass from which the force is thought to emanate. The center of the other mass affected by the force would be understood to lie upon the surface of the sphere in which the effecting mass is centered.

Though the force of gravity would thus appear to be fixed in its proportions, it is difficult to speak of its increase or decrease at varying distances without incorporating a sense of the passage of time. For there is a proportionate diminution of the force as the distance between bodies increases. And the reverse occurs when the distance decreases. Moreover, these increments of proportional change can be made as small as the calculation of an instant of time will allow.

Yet the pervasiveness of this force is immediate. It does not travel to its destination. It is instantaneous action at a distance. It is as though gravitation were, in fact, not a form of energy unfolding, but of energy unfolded. It is as if time stood still. Thus it is what would be seen if time did not exist and all the

The Limits of Reason

changes in the extensions of space were wrought simultaneously.

This is Newton's gravitational force, but not the action of that force. The dropping of a spoon or the action of the moon in orbiting the earth are not being spoken of here. These are the events, or changes, explained by gravitational force. But they do not explain the force itself. Such a force must be conceived statically as an imaginative insight: a picture for organizing thought. Thus there is no reason to assume the actual existence of such a force. What exist are the proportional relations found in experience and the changes wrought in terms of them.

George Lowell Tollefson

Zeno's Paradox

Only one of Zeno of Elea's four paradoxes concerning motion will be discussed, since the others entail similar problems. The Achilles paradox develops its argument on the basis of a fallacy, which is time conceived as an entity possessing a separate status from space. It begins with a race between Achilles and a tortoise. Since the tortoise is slower than Achilles, it is allowed to start first. Presumably Achilles can never catch up with the tortoise because, whenever he reaches a point where the tortoise has been, the tortoise has moved on.

Achilles draws ever closer, as the distance between him and the tortoise becomes shorter. But there is always some fraction of a distance he must traverse, as the tortoise simultaneously moves on beyond that point in space towards which Achilles is striving. So, since the diminishing increments of distance between Achilles and the tortoise are infinite, and the increasingly shortened moments in time required to traverse them are also infinite,[46] Achilles never catches up to, let alone overtakes, the tortoise.

The fallacy here lies in thinking of increments of time as possessing a separate status from increments of space. Moments, seconds, and minutes are supposed to be increments of time, as though time were an extended entity. But time is not

---

[46] Again, a caveat must be acknowledged concerning the present author's discomfort with the term "infinite" as applying to material entities. The preferred term is "indeterminate," as in "indeterminately small." However, in the various renditions of the paradox, the term "infinite" is generally used. So it is used here.

physical and therefore not extended. And there can be no increments of something which is not extended. So, to compensate for this shortcoming, moments, seconds, and minutes are based on the physical increments of extension which are observed in physical change.

Objects which are assumed to occupy space are extended. So, in a simple quantitative description, space can be measured in terms of the extended objects within it. It can be measured in physical increments of distance. Whereas, although language appears to have supplied time with similar but independent increments—referred to as moments, seconds, minutes, etc.—time does not truly possess its own increments.

Rather, time exists for human awareness only because human beings observe changes in the distances of space. For change, even the subtlest imaginable change, like physical or chemical change, involves an alteration in the mutual distance relationships of the parts or elements involved.

In other words, when an extension of space—an object or a chemical compound—is altered by change, such an alteration involves a change in location, or spatial distance, however minute. For, if a part is altered in its relation to other parts, or a compound is replaced by its elements, that is a displacement.

Even nothing more than a change in color or the alteration of a shadow is experienced as though it were a displacement of one property of an object by another. It is as if the color or shadow moved. And another took its place. In the enlarged and more obvious case of discernible motion, like a moving car, one object changes in its distance relationship to other objects. The other objects are the background against which the car is moving.

Now the time which measures these changes is itself measured by means of physical changes. For example, let it be measured by a mechanical or digital clock. The changes in the instrument involve observations of alterations in physical distance. These may be the movement of hands on the face of the clock or the digital reconfiguration of its numbers.[47] Whatever the case may be, displacement occurs. This displacement involves a restructuring of distance relationships either between two different positions of the hands on the face of a clock or between the reconfigured elements of the digital display of the numbers.

Thus time is distance measured against distance: one position or configuration as opposed to another, as already explained earlier in this work. It is the distance a second hand is from where it previously was in relation to a clock face. Or it is the realignment of digital elements on the clock face, resulting in altered distances among the elements. Thus an incremental unit of time is a change in physical distance.

These changes in distance are made uniform by convention, so that all increments of time, except moments, are proportionally in agreement. One type of increment is a second. Sixty of these is a minute. Et cetera. Thus each in its kind becomes part of a uniform standard of measure for time. They are then used to demark other physical changes in distance. So, when a person speaks of the time it takes for something or someone to traverse a few inches or a few yards along a given track, it is understood by this sort of common measure.

---

[47] Such a reference to an alteration in distances can even be applied to the workings of an atomic clock. But the explanation is more complex.

The Limits of Reason

In this way, time is inextricably linked with spatial references. For it is a spatial reference applied to other spatial references. Thus it can be seen that time and the distances Achilles and the tortoise travel are all three founded on measures of distance. However, insofar as time is concerned, moments appear to be an exception to incremental uniformity in that they are variable. They are nevertheless variable in a physical manner. They can be divided into greater or lesser increments involving a physical reference just as the track traversed by Achilles and the tortoise can be divided arbitrarily into varying parts of lesser or greater length.

It may be asked how the motion of Achilles and the motion of the tortoise might be compared, or brought into a relationship with one another. This is accomplished by means of their differing progress along the same track. These differing progressions are conceived as rates of change in distance.

So they are rates of change in distance covered along the track. And they involve the time in which each change in distance takes place. The time is measured as the distance a hand moves on a clock. Thus Achilles and the tortoise each traverse a certain distance on the track in accordance with the distance a hand moves on a clock.

Accordingly, it is the distance the hand moves on the clock which is the single comparative reference linking the distances Achilles and the tortoise move on the track. In other words, Achilles and the tortoise may cover different distances per unit of time. But the units of time applied to each are the same. And they are themselves measures of distance.

So, in reference to both Achilles and the tortoise, a single standard of distance for measuring other distances is being used. The distances Achilles and the tortoise move on the track are also linked by the fact that they represent percentages of the coverage of a common track. For Achilles and the tortoise are competing over the same total distance.

To illustrate the issue, let it be assumed that Achilles moves a distance of three yards in the time it takes for the tortoise to move a distance of three inches. These are their differing rates of progress. To express this comparison as "in the time it takes" is to assert that the distance of three yards is linked to the distance of three inches by the movement of a second hand over a few millimeters of a clockface. This is the time it takes for both distances to be simultaneously traversed.

So now let their mutual progress be followed from the beginning. The first movement of the second hand belongs to the tortoise alone. The tortoise moves three inches. The second movement of the second hand belongs to both Achilles and the tortoise. Achilles moves three yards. And the tortoise moves another three inches. So, by a common measure of two movements of the second hand, they move three yards and six inches respectively.

Thus in one double movement of the second hand, Achilles has moved three yards to the tortoise's movement of six inches. These movements, measured in the same way, are on the same track. So Achilles catches up to the tortoise and passes it by. For his movement covers a greater percentage of the common track.

There is no mysterious dimension of time involved. For there is no standard of measure composed of unique incremental elements independent of spatial distance. There is only spatial distance, whether one speaks of the passage of time or of the movements of Achilles and the tortoise. For time as a means of incremental measure functions as no more than a literal yard-

The Limits of Reason

stick held up to the actions of Achilles and the tortoise so as to create a standard of comparison between them.

The "yardstick" of time is a distance measured by the progress of a hand on the face of a clock. It is a third reference in distance linking the two distances of Achilles and the tortoise as they traverse the same track. The second hand moving on the face of the clock is not something different in kind from what occurs on the track. It is a distance as well. So it does not exhibit a categorically independent existence. For the face of the clock is an extended surface in physical space, just as the track is.

The track is a distance and no more. The movements of Achilles and the tortoise are distances and no more, integrated into the same track. That portion of the face of the clock which is traversed by the second hand is a distance and no more. It is compared with the distances on the track as a common external reference for them. For it occurs in simultaneity to them.

But it is still a distance. Thus it does not exhibit an independent existence as a mysterious dimension, alienating it from physical distance in such a manner that it can be used to draw out the process of the race and divide its progress into infinite independent steps. For the steps will be lengthened or shortened in proportion to *its* being lengthened or shortened.

If it should be argued that another measure is needed to account for the simultaneity between the above three distances, that measure will also involve a change among extensions in physical space. And that change in extensions will involve some sort of rearrangement in their mutual positions. This rearrangement of mutual positions is marked by a physical distance. So on ad infinitum.

George Lowell Tollefson

Momentum and Location

Today any concern in quantum mechanics regarding the fact that both the momentum and location of a subatomic particle may not be determined simultaneously is more likely to raise philosophical issues than technical ones. Nevertheless, a closer examination of the concept of motion is warranted.

Let an analysis of the problem of determining the locality of any object which is in motion be brought under review. Take a projectile. For the sake of convenience, imagine that it travels its own length in one second's time. It would then travel five times its length in five seconds.

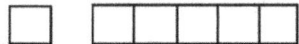

As shown in the figure, the projectile can be represented by a single independent square. And the distance traveled in five seconds can be illustrated with a rectangle divided into five contiguous squares, each the size of a single square like that of the projectile.

Looking at this simple situation, it might be assumed that at the end of three seconds the projectile would have traveled a distance of three squares and would be securely located within the boundaries of the third square. But would this be true? As has been previously noted, since the projectile is continuously moving, it must be asked: at what precise moment would the projectile's dimensions be coterminous with the third square?

The Limits of Reason

There is no time interval within which this could be determined. If a high speed camera were imagined to stop the movement of the projectile when it appeared to be coterminous with the third square, that would only be an approximation: a crude representation in the form of an image. A much faster camera would reveal this.

So it is evident that, if the time intervals were divided into the smallest conceivable instants, and the movement of the projectile likewise limited to the smallest spatial advances, the projectile would still be found to be moving. What this demonstrates is that a momentum and a location cannot be simultaneously determined for the projectile because motion is, by definition, change of place. So the emphasis is on change, not place.

These observations concerning the character of change have already been mentioned. But in practical circumstances, particularly in macrophysics, there is little reason to be concerned with them. For the conceptual conflict does not occur in ordinary calculation at the macrophysical level. This is because there is a limit on how much precision is needed. And the level of that limit is far below any macrophysical demands. Hence the calculation of an instantaneous velocity has many practical applications, though there is no such thing as an instant.

But, when a move to the microphysical domain is made, there are difficulties in experimental observation. So they result in a greater focus of attention upon the details of what is occurring. Nevertheless, the problem is essentially conceptual, not practical. It is this: when momentum and location are compared in thought, their interrelationship becomes confused.

This is because the two concepts sufficiently overlap in meaning to highlight an inconsistency between them. Both concepts are founded upon an idea of location. Momentum places its emphasis upon change in location. And, as the change is continuous, it cannot specify position. But, when understood independently of momentum, the concept of location does specify position, consequently eliminating the process of change which is characteristic of momentum.

The point being made is that the problem is not so much physical as conceptual. It asks how the observed phenomena should be conceived. But, for practical purposes, these philosophical considerations can be ignored, since they do not impact the usefulness of the physical calculations. For physical science is only concerned with what can be observed. It is in aid of this that its concepts are established. Thus it is satisfied not to go beyond the results of verified observation.

So it is from a purely conceptual perspective that an issue arises. For it is the concepts which are in conflict. They are initially thought to relate ideas which are not mutually contradictory in meaning. But such subtly integrated concepts do contradict one another. Treating them as though this is not the case creates confusion.

Universal Complementarity

Of nothing whatsoever can the motion and location be known simultaneously. For, if things should be made to appear otherwise, it is the result of a confusion of expression or an illusion of theoretical practice. An object moving freely in space has already been imagined. So let another example be considered.

What would be the character of an attached object moving along an arc? An illustration of this would be a pendulum. At various times, it may be set in motion. But it cannot be moving and in a state of rest simultaneously. It must be in one or the other of these two states. The alternative to the state of motion is rest, and vice versa.

If the bob of a pendulum reaches the end of its arc and reverses motion, it cannot be determined that it was in a state of rest in the fleeting moment of the reversal. This is because its motion diminishes incrementally toward a vanishing point in one direction and increases in the same manner in the other. In what time interval is the motion reversed?

Where there is motion, the emphasis is upon process. In the case of rest, the emphasis is upon the absence of process. In the former case, the bob is moving. In the latter, it is at rest. In a state of rest, there is no alteration in location. For it is duration in a particular location which determines rest. Likewise, rest determines location.

Movement abrogates location by continuously changing it. Consequently, it is the inability to fix a location which indicates motion. Motion occurs within a time interval. And change is the register of time. Conversely, the specification of a location indicates a lack of motion. This too occurs within a time interval, which is measured by the relative motion of something else.

Like motion, a state of rest involves both space and time. For location is not only a spatial condition. In reference to an object in a state of rest, it is temporal as well. The specification of a location implies the duration of a state of rest. Thus, on the one hand, in reference to a state of motion, both time and location are involved in its process of change. But, in regard to a state of rest, time continues elsewhere in a state of change, while the spatial relations of the object indicate a state of rest. They are fixed, until motion is transferred to the object again.

Consequently, in addition to the passage of time, a location is required as an indication of a state of rest. If the location of an object remains the same for any interval of time, however minute the interval, the object is at rest for that interval. Motion, on the other hand, involves a progressively sliding scale of locations, such that no one location is more applicable to the object in motion than any other. For in motion there is no duration of location. There is only duration of change in location.

Neither does this inapplicability of location depend upon variations of momentum. So long as there is a change in location caused by motion, regardless of how gradual or varied the change, and even if the motion appears not to be taking place at all, the duration of any one location cannot be brought within reach of the moving object. For what arrives within the environs of a location, much less some part of what arrives there, is passing away. Thus location is continually changing character.

In other words, the essential property of motion is that it involves an instability of location, without any possibility of the characteristics of any one location being definitively identified

The Limits of Reason

with the moving object while the motion continues. The motion contains no states of rest. Consequently, it is something entirely different from rest. It is a series of changes, or a process. Rest, on the other hand, is devoid of change. It is the absence of process. So, for a state of rest to be determined, time must pass, as determined by events external to the state of rest. But location does not change.

This is to say that change does not occur in the state of rest. For change involves displacement, however subtle. But, in spite of this condition of quiescence, the measure of time's general expenditure is found elsewhere, where change is occurring. For example, an apple is at rest upon a table. But nearby a clock is ticking.

The principal point of the argument presented here, as well as in the preceding essay, is that, under the physical and temporal conditions of motion, there can be no such thing as a determination of the position of a moving object, regardless of the type of movement. And neither can time stand still. For it is a characteristic of motion, which is change. For this reason, it may be the case that such a concept as instantaneous velocity can exhibit practical utility to a high degree of precision. But it does not accord with the physical and temporal facts of experience.

George Lowell Tollefson

The Hidden Dynamic

Energy is a word designating proportional relations. In other words, it is a word designating mathematical relations, which in turn express change in quantitative terms. It explains nothing beyond this. Thus a person is forced to take any further investigation of the matter outside the material realm.

That is, he is required to look at something other than physical experience, which gives him extensions and their proportional relations without explaining the changes that occur among them. So he looks beyond material experience for some kind of an accounting of change. He looks to the circumstances of his awareness.

But awareness is consciousness. And unfortunately they mean the same thing. Moreover, consciousness reveals nothing about material experience but its order. That order involves motion, or change. This leaves human understanding in a convoluted, even circular condition, since it would appear that motion causes motion, or change causes change.

This circularity is the essence of causal relationships in physical experience. It does not explain why they occur. So an awkward attribution to consciousness of characteristics borrowed from human experience becomes necessary. For the limitations of human understanding leave no other alternative.

Therefore, the word "dynamic" will be used as a synonym for "consciousness." "Dynamic" is a useful metaphor for describing the domain of the potential, as opposed to the purely actual, since change, in coming into being, expresses a potential. Moreover, anything dynamic involves change, as in a

The Limits of Reason

change from the potential to the actual. Thus consciousness is the domain of potential. It is apparently from this that change proceeds.

Furthermore, not wishing to assume that consciousness is a black hole of creativity, inexplicably spitting out raw experience without purpose, seems justified. For, if such a circumstance is assumed, there can be no hope for an explanation of change, since any explanation implies an assumption of order, and order an indication of purpose. So it is reasonable to presume that consciousness and its conversion of the potential into the actual exhibits purpose.

Now this hidden dynamic in consciousness becomes a metaphor indicating an inconceivable complexity, since its workings are presumed to be determined in some manner known only to it. That is, they cannot be identified by human understanding. In other words, a human conception of the causal origins of experience in consciousness cannot be formed. Whether or not the activity within consciousness resembles what is thought of as causation is not known. Therefore, for human understanding, mystery is its character.

Human experience, on the other hand, is the domain of what exists for human sensibility. It is that content of consciousness which is present to human awareness. It is the material realm of the actual. In contrast, consciousness is the domain of the potential. Thus the potential does not exist for human sensibility, except as the ground of that sensibility.

So the fact is that there can be no empirical acquaintance with the origin of change in human awareness other than that it arises in consciousness. Nothing is known about it beyond speculation. For this reason, the concept of a potential, when understood in these terms, is no more than an *implication* of change.

But that which is actual within consciousness, while potential for human awareness, does become actual for human awareness. As it does, it enters human experience as a disturbance of the content of consciousness. A human mind can go no further with this explanation. For, beyond giving the concept of potential this definition, what it is is unknown. However, when understood in terms of human experience, the potential can be viewed as the unfolding of the actual through change.

The actual, surveyed strictly in terms of itself, is not at all like the potential. For it carries within its nature nothing of the latter. It is merely actual. As such, it may be conceived much in the way that past events are conceived. As soon as anything becomes actual, it is past. Nevertheless, though the realm of human experience is not the origin of change, it is where change takes place.

Change is motion, as has already been stated. Even chemical change is motion. Thus, as understood strictly in terms of human experience and its ordered character, change is an intrusion upon the equilibrium of order. So, were there no dynamic from within consciousness acting upon that order, change would be inconceivable and nonexistent.

When examined within the limits of human awareness, an object in motion always appears to be moving through a series of increments of space. This is a series of locations. It is how motion is experienced, no matter how minute the extension of space or the interval of time during which an attempt might be made to locate the moving object. The moving object is always in a state of transition, always passing through a set of locations, never definitively in any one location. It is never deter-

The Limits of Reason

minedly fixed as to place. Thus motion does not participate in the static character of a given order.

So how can its introduction into experience be understood? How can its observable behavior there be accounted for? Let it be described using an image of four-dimensional space-time. Four dimensional space-time would not, in this case, be conceived as dynamic. Nor can it be represented in the mind. For such a schema does not exist for human perception.

Rather, it may be said that change acts in such a manner as to fold this four-dimensional space-time into the three dimensions of space which the human mind can be aware of. How does it do this? Through the working of time. For time is the measure of change. Thus time—an essential ingredient of change when change is understood as motion—is withdrawn from the suggested four-dimensional space-time configuration.

As a result, a sense of time unfolding in three-dimensional space emerges. This is the manner in which the underlying character of potential is revealed to human sensibility. It is revealed as change unfolding through time. But, of course, it is the change itself, the motion, which is the measure of time.

This is the reason why the human mind cannot discern a moving object's position. For the ongoing change described by a moving object is an expression of potential. Potential is what is not yet actual. In becoming actual, it is either an adding or a withdrawal of extensions. It becomes a reworking of three-dimensional space.

This is either a modification of the location of extensions in relation to one another, as in the changing relationship between a moving object and its background. Or, as regards change within a single extension, it is a working of new arrangements within that extension. Thus these varying states of extensions are an unfolding into actuality. But, since in the state of being potential the elements of change are not actual, they cannot be registered in human awareness. So what is experienced is change without an initial cause.

The Limits of Reason

Instrumental Science

In criticisms of the concept of energy contained in this work, its practical applications in physical science have not been questioned. Rather, a distinction between functionality and philosophy has been the goal. Energy is a useful functional concept. But, at best, it fits awkwardly within a more inclusive construct of human experience. So it may sometimes be necessary to make the distinction.

In fact, it may, for broader philosophical purposes, be necessary to create a more universal template for reality, one that does not express the concept of energy. Doing this could help to resolve certain anomalies within general thought which the idea of energy creates. There has been an attempt to demonstrate this throughout the latter section of the present work.

Nevertheless, this philosophical viewpoint is not fit for all occasions. So it is certainly understood that, for the sake of practical application within physical science, the concept of energy will apply. And, for the sake of that practical utility, it can be given universal status, when that status is meant to reference a physical universe only. In short, the concept of energy not only is, but ought to be, a part of any quantitative explanation of physical reality.

But when the whole of human experience is taken into account, a distinction should be drawn. This is because science is not about what is true and what is not. It is about what relates human beings in an instrumental way to their physical or biological world. This involves observing and measuring one concrete set of physical phenomena in terms of another, or even conceiving—that is, inventing—such phenomena where observation is rendered obscure and indirect.

So, for the sake of a broader philosophical unity in thought, there are times when it is necessary to express the whole of human experience rather than a subclassification of the same, such as physics, chemistry, or biology reveal themselves to be. For human beings must relate not only to particular subclassifications within the greater whole of experience, but to that greater whole as well. Of necessity, this is done in different ways, as there is more than one philosophical approach. So any human access to knowledge and understanding requires a flexibility of thought. Such flexibility precludes a monolithic approach.

Final Cause

Modern physics is moving in the same direction as biology—that is, toward final cause explanations of phenomena. Biology moved into this realm with evolutionary theory. Because it was dealing with an interrelationship between complex systems, it could not explain things in terms of efficient cause. A person could not explain why a polar bear had a white coat by searching out a particular efficient cause or set of such causes.

She had to consider the polar environment, a system far too complex for singling out efficient causes. Taking the polar environment into consideration, it could then be seen how the bear (also a complex system) fit into this environment. She could determine the bear's function, or purpose, in that environment and thus the function, or purpose, of its white coat.

It can be assumed that efficient causes have played a hidden, multiplex role in the development of the white coat. But complexity or inaccessibility of such causes prevents an articulation on their behalf. Thus a reliance upon a final cause explanation is necessary. For it subsumes multiple efficient causes under a general system and articulates the overall system in its relationship to another system. The articulation is made in terms of purpose.

This was a mode of thinking expressed by Aristotle but set aside during the development of modern science in the sixteenth and seventeenth centuries. However, it was reintroduced by Charles Darwin in his *Origin of Species*. In spite of the later development of Neo-Darwinism and its reliance upon Mendelian genetics and a kind of reductionism in favor of physical mechanics, the principle still holds.

But one might ask, how is it that modern physics may be said to be moving in this direction? Let it be assumed that the effort to search outward into the unlimited reaches of the cosmos closely resembles a similar effort to probe inward into the submicroscopic relations of the subatomic world. Both directions of enquiry encounter increasing difficulty as they move further from the medial realm of the everyday world of human experience.

Soon, in pursuit of such enquiries, a person finds herself at a point in the cosmos where observations are obscured by a compounding of inference, or at a point in the subatomic realm where accustomed causal relations do not seem to hold. In the former case, she is threatened by an increasing uncertainty in her deductions. In the latter, she accepts the apparent anomalies and entertains a probabilistic approach as a means of relating them.

Either way, she begins to treat individual causes as unapproachable and to group phenomena into complexes, or systems, both mathematical and physical. In other words, what is being said is that individual effects cannot rest upon single instances of efficient cause. So an observer begins to approach them within a larger intellectual framework.

Looking at it in this way—that of grouping phenomena into systems, rather than limiting consideration to individual causes—a method emerges of generalizing and systemizing data. The concept of entropy, as applied to the behavior of gasses and to energy dissipation in the universe, reflects this approach.

A full recognition of the fact that this is final-cause thinking has not been established in physics. Nor has it been openly ac-

The Limits of Reason

cepted in biology, for that matter. But the use of probability and such concepts as complementarity, uncertainty, and superposition in the quantum world are a confession of some loss of control over efficient cause. On the cosmic scale, the fact that universal entropy cannot be demonstrated is a similar instance.

This is also why what the polar bear's adaptations were going to be could not have been precisely stated before they were encountered. Of course, the use of probability in understanding biological variation or in mentally organizing events in the quantum world implies that a fuller, more complete accounting of any set of phenomena under study ought to be both expected and desired.

But the search for knowledge has become increasingly subtle and indefinite. Therefore, general assumptions will continue to aid the human mind in its grasp of the very large, the very complex, and the very small. This is an indication of human limitation and not a fundamental characteristic of reality.

George Lowell Tollefson

The Discrete Mind

As a matter of necessity and convenience, human beings have long grown accustomed to understanding material experience in discontinuous terms. For their ordinary concepts of the world are built around discrete entities. Physical science, on the other hand, is increasingly coming to understand that the material world is a continuum.

There are no discrete entities in material experience. Conversely, images of perception, as well as the conceptual formulations of the human mind, focus upon isolations of phenomena. Thus they are necessarily discrete. So, however hesitantly accepted this insight may be at the level of common experience, there are no discrete entities in material experience prior to the intervention of the mind.

This view is supported both by the structure of mathematics and by the dependence of physical science on a mathematical basis. Mathematics bridges a gap between the discrete and the continuous. For, though it appears to be discrete in the character of its numerical distinctions, it is built upon a foundation of indiscernibles. These indiscernibles reflect an underlying continuity.

The continuity is found in an exact similarity between arithmetical units. Arithmetical units are an abstraction, having no properties in themselves but extension. So, aside from their functional employment, which may involve variations in magnitude, they are indiscernible one in comparison to the other.

And where this is the case, there is no distinction. Where there is no distinction, there is continuity. However, there is a

The Limits of Reason

distinction in numbers because there is a differentiation between sets of arithmetical units in the numbers. For numbers are composed of these units. And mathematics differentiates between them, as in the distinction between the sets of units in 2 and 3.

So the continuity is in the units. The distinction is in the numbers. This development of distinctions upon a foundation of continuity is what informs the character of discreteness in mathematics. A continuity of underlying indiscernibles is necessary to support the unity of its differentiated numerical concepts. For, without this underlying unity, it could not be a logically integrated system. The intellectual construction of such a system is a convenience of the mind, which was adopted in recognition of the human need to operate among related but distinguishable entities.

But to obviate problems that a system of discrete numbers inevitably creates, incommensurables were investigated, irrational numbers discovered, and differential and integral calculus developed. These press their meaning and practice in the direction of the fundamental continuity of material experience. For they transcend the discreteness of numbers.

So, in its recognition not only of difference and discreteness, but of continuity as well, mathematics has been developed into a bridge between the practical necessities of human existence and the fundamental nature of things. It stands as humankind's most remarkable compromise between the mind and its material experience. It is an effective logical paradigm designed to establish a workable link between the things people wish to know and the limits in the human capacity for knowledge.

What this means is that all creations of the human mind are limited and discrete. They are so insofar as they can be conceptualized without being expressed in terms of a negative reference. It is for this reason that rational numbers are conceived in terms of themselves, while irrational numbers must be conceived simply as not rational.

The former are limited and discontinuous. Whereas the latter are unlimited and form a continuum. Consequently, rational numbers express the mind's quantitatively discrete and limited organization of material experience. In opposition to these stand the irrational numbers, which exhibit the character of that experience prior to the mind's organization of it.

The Limits of Reason

Index of Names

Archimedes, 5, 5 (n. 4), 46–47, 46 (n. 19), 50
Aristotle, 210
Brahe, Tycho, 182
Cantor, 153
Copernicus, 182
Darwin, 210
Dedekind, 140, 140 (n. 38)
Descartes (Cartesian), 14, 70
Euclid, 1, 2, 2 (n. 1), 3 (n. 2–3), 4–8, 7 (n. 6), 9 (n. 7), 11, 11 (n. 9), 13, 16–19, 22, 22 (n. 11), 23 (n.12), 24, 24 (n. 13), 26, 26 (n.14), 28–31, 29 (n. 15), 30 (n. 16), 32, 33, 37–38, 49–50, 49 (n. 20–21), 56, 58–61, 58 (n. 22), 59 (n. 23–24), 60 (n. 25), 63–64, 64 (n. 26), 66–71, 68 (n. 27), 70 (n. 28), 71 (n. 29–30), 99, 104–105, 104 (n. 31), 107, 122–124, 123 (n. 33), 124 (n. 34), 132, 133 (n. 36), 135–137, 139–140, 140 (n. 39)
Galileo, 182
Heisenberg, 180 (n. 43)
Kant, 149
Kepler, 182
Lagrange, 182
Laplace, 182
Mendel (Mendelian), 210
Newton, 182, 188, 190
Ptolemy (Ptolemaic), 181
Pythagoras (Pythagorean), 9, 35, 48
Russell, 153 (n. 40), 156, 157 (n. 41)

The Limits of Reason

Thomson, Sir J. J., 20
Zeno of Elea, 191

… George Lowell Tollefson

Bibliography

Archimedes. "Measurement of a Circle." In *Euclid, Archimedes, Apollonius of Perga, Nicomachus, Vol. 11, Great Books of the Western World*, translated by Sir Thomas L. Heath. Chicago: Encyclopedia Britannica, Inc., 1952.

Aristotle. "Physics." In *Aristotle I, Vol. 8, Great Books of the Western World*, translated by R. P. Hardie and R. K. Gaye. Chicago: Encyclopedia Britannica, Inc., 1952.

Darwin, Charles. "The Origin of Species." In *Darwin, Vol. 49, Great Books of the Western World*. Chicago: Encyclopedia Britannica, Inc., 1952.

Dedekind, Richard. "Continuity and Irrational Numbers." In *Essays on the Theory of Numbers*, translated by W. W. Beman. New York: Dover Publications, Inc., 1963.

Euclid. *Euclid's Elements*, translated by Thomas L. Heath. Santa Fe: Green Lion Press, 2010.

Heisenberg, Werner. *Physics and Philosophy*. New York: Harper and Row, Inc., 1962.

Kant, Immanuel. "The Critique of Pure Reason." In *Kant, Vol. 42, Great Books of the Western World*, translated by J. M. D. Meiklejohn. Chicago: Encyclopedia Britannica, Inc., 1952.

Newton, Isaac. "Mathematical Principles of Natural Philosophy." In *Newton, Huygens, Vol. 34, Great Books of the Western World*. Chicago: Encyclopedia Britannica, Inc., 1952.

Russell, Bertrand. *Introduction to Mathematical Philosophy*. www.digireads.com: Digireads.com Publishing, 2010.

———. "Mathematics and the Metaphysicians." In *The World of Mathematics, Vol. 3*. New York: Simon and Schuster, 1956.

Tollefson, George Lowell. *The Immaterial Structure of Human Experience*. Santa Fe: Palo Flechado Press, 2019.

Waterfield, Robin. "Zeno of Elea." In *The First Philosophers*. New York: Oxford University Press, 2000.

www.ingramcontent.com/pod-product-compliance
Lightning Source LLC
Chambersburg PA
CBHW072003110526
44592CB00012B/1187